NONTOPOLOGICAL SOLITONS

World Scientific Lecture Notes in Physics Vol. 24

NONTOPOLOGICAL SOLITONS

Lawrence Wilets

University of Washington
Seattle

World Scientific
Singapore • New Jersey • London • Hong Kong

Published by

World Scientific Publishing Co. Pte. Ltd.
P O Box 128, Farrer Road, Singapore 9128

USA office: World Scientific Publishing Co., Inc.
687 Hartwell Street, Teaneck, NJ 07666, USA

UK office: World Scientific Publishing Co. Pte. Ltd.
73 Lynton Mead, Totteridge, London N20 8DH, England

NONTOPOLOGICAL SOLITONS

ISBN 9971-50-697-1

Printed in Singapore by JBW Printers & Binders Pte. Ltd.

Dedicated to

T. D. Lee

and to

my collaborators who made this book possible

PREFACE

For more than seven years I have been heavily involved in various aspects of nontopological solitons, particularly those inspired by the Friedberg-Lee model of quarks, gluons and scalar fields. During that period, I have been fortunate to have a number of superb collaborators with whom I shared the joys and frustrations of discovery, and from whom I learned as well as, I hope, contributed. I have borrowed substantially from those collaborations in preparing this monograph. I have frequently paraphrased or quoted from our publications. Therefore, let me begin by expressing my grateful thanks to Drs. J. Achtzehnter, M. Bickeböller, M. C. Birse, J-L. Dethier, G. Fai, R. Goldflam, E. M. Henley, R. Horn, G. Krein, M. Li, E. G. Lübeck, H. Marschall, R. Perry, H. Pirner, J. J. Rehr, A. Schuh, F. Stancu, P. Tang and A. G. Williams. Special thanks are due Drs. Stancu and Williams for a thorough reading of the complete manuscript, providing final corrections and suggestions. I am indebted to Prof. W.-Y. Pauchy Hwang for introducing me to soliton models. I also wish to acknowledge the kind hospitality of the Lawrence Berkeley Laboratory and the Stanford Linear Accelerator Center, where I spent my sabbatical leave during 1987-88 when this book was written.

Perhaps the simple title of this work is actually pretentious, for I have not attempted to cover the broad field of nontopological solitons. I apologize to those whose work I may have slighted. Since by definition a monograph covers a particular subject in depth, I have chosen to concentrate much of the material on those topics with which I am most familiar by virtue of my own involvement, namely Friedberg-Lee type solitons bags. Nevertheless, I hope that this work will find use beyond the specific applications presented here, for much of the methodology is relevant to a variety of problems in relativistic field theories and collective dynamics. I would like to call particular attention to the sections dealing with coherent states, projection and boost, generator coordinates, and gluon (Maxwell) propagators in a dielectric media.

CONTENTS

Chapter one

INTRODUCTION

1.1 Why Quantum Chromodynamics?

It is an article of faith that quantum chromodynamics (QCD) is the fundamental theory of the strong interactions. At high energies and large momentum transfer, where asymptotic freedom validates perturbation theory, there are impressive successes. From such experiments can be deduced the generations of quarks, their masses, charges and quantum numbers.

It is in the realm of low energy (GeV), large scale (fm) physics that the predictive power of QCD falters. The nonlinearity of the theory has presented theorists with a formidable task which has not yet yielded to analytical treatment, although vital qualitative features have emerged. On the other hand, lattice gauge calculations provide a nonperturbative numerical approach which has long held the promise of providing reliable, detailed solutions. Unfortunately, lattice calculations have proved to be extremely expensive in computation time and the results, to date, have been of limited accuracy and utility. Even the strength of the linear component of the confinement potential is not cleanly predicted, since it has not yet been possible to carry the calculations into the region of linearity; the linear component can only be extracted by assuming an explicit form for the potential and then fitting the calculations to that form. The parameters of the assumed confinement potential can be extracted empirically from heavy quark spectroscopy, and the identification of the strength of the linear term is often used to set the scale of lattice calculations. There is every reason to expect further improvements in lattice calculations as computer facilities expand, as the inclusion of dynamical quarks becomes more reliable, and more extensive analysis of the "data" becomes available.

QCD is part of the standard model, which incorporates the leptons and fields of electroweak interactions. There have been as yet no experimental violations of the standard model, but there do remain theoretical problems, and it is certainly incomplete. Among other things, it gives no mechanism for determining the various quark and lepton masses, the mixing phase angles, nor the couplings to the Bose fields. Solutions to these problems are among the goals of superstring theory. Here we will restrict ourselves to the modeling of QCD.

Why QCD? Because *it is the only game in town.*

1.2 The QCD Lagrangian

QCD is the marriage of two concepts in particle physics: the quark (parton) model, and the theory of local gauge fields. Quarks were introduced by Gell-Mann (1964) and Zweig (1964) as a purely mathematical construct to provide a framework for the SU(3) classification of elementary particles which in turn was an outgrowth of the eight-fold way. In spite of early resistance, experiments bestowed on the objects a reality of their own, and this in turn demanded the addition of a new quantum number, color, to conform to the Pauli exclusion principle. The development of Yang-Mills (1954) theory, based on the requirement of local gauge invariance, led to the introduction of gauge fields—gluons—which became the force fields, interacting with Fermions and among themselves.

The basic QCD Lagrangian density is

$$\mathcal{L} = \overline{\psi}(i\gamma^\mu D_\mu - m)\psi - \tfrac{1}{4}F^c_{\mu\nu}F^{\mu\nu}_c \tag{1.1}$$

plus Higgs bosons and appropriate counter terms. The summation convention for repeated upper-lower indices of any kind is assumed. The various constituents have the following meaning:

The quark mass m is a diagonal matrix. Since free quarks are not observed, one can only speak of quark masses in regions of confinement or asymptotic freedom. *Current* quark masses are used in relativistic bag and soliton calculations; *constituent* quark masses are used in nonrelativistic potential models. Current masses for the light, u and d, quarks are very small, while the corresponding constituent masses are roughly one-third of a nucleon mass. The larger constituent mass can be ascribed to the energy of confinement. For example, a massless quark in an MIT bag of radius R is rather like a nonrelativistic quark of mass $2.043/R$, which is the energy of the lowest bag eigenvalue. Clearly any quoted quark masses are model dependent. Table 1.1 gives approximate quark masses along with other properties.

Throughout this book, we follow the notation of Bjorken and Drell (1964). We use units where $\hbar = c = 1$. Energy then has units of inverse length and the units are sometimes quoted in MeV or GeV and sometimes in fm^{-1}; lengths are usually quoted in fm, but sometimes in MeV^{-1} or GeV^{-1}. $1 \text{ fm}^{-1} = 197.33$ MeV.

Table 1.1. Properties of quarks. Masses are model-dependent and hence have only qualitative meaning. Mass units are MeV. (*cf.* Gasser and Leutweyler, 1982).

flavor	symbol	charge	current mass	constituent mass
up	u	$+\frac{2}{3}$	6	330
down	d	$-\frac{1}{3}$	10	330
charmed	c	$+\frac{2}{3}$	1,350	2,000
strange	s	$-\frac{1}{3}$	199	520
top ?	t	$+\frac{2}{3}$	>50,000	>50,000
bottom	b	$-\frac{1}{3}$	5,000	5,000

The field operator ψ is taken to be a vector incorporating 6(?) flavor and 3 color, as well as the four Dirac, components. D_μ is the gauge-covariant derivative

$$D_\mu = \partial_\mu - i\tfrac{1}{2}g_s\lambda_c A_\mu^c. \qquad (1.2)$$

The λ_c are the $3\times 3 - 1 = 8$ traceless Gell-Mann $SU(3)$ color matrices, constructed in analogy with the Pauli spin matrices:

$$\lambda_1 = \begin{pmatrix} 0 & 1 & 0 \\ 1 & 0 & 0 \\ 0 & 0 & 0 \end{pmatrix} \qquad \lambda_2 = \begin{pmatrix} 0 & -i & 0 \\ i & 0 & 0 \\ 0 & 0 & 0 \end{pmatrix}$$

$$\lambda_3 = \begin{pmatrix} 1 & 0 & 0 \\ 0 & -1 & 0 \\ 0 & 0 & 0 \end{pmatrix} \qquad \lambda_4 = \begin{pmatrix} 0 & 0 & 1 \\ 0 & 0 & 0 \\ 1 & 0 & 0 \end{pmatrix}$$

$$\lambda_5 = \begin{pmatrix} 0 & 0 & -i \\ 0 & 0 & 0 \\ i & 0 & 0 \end{pmatrix} \qquad \lambda_6 = \begin{pmatrix} 0 & 0 & 0 \\ 0 & 0 & 1 \\ 0 & 1 & 0 \end{pmatrix}$$

$$\lambda_7 = \begin{pmatrix} 0 & 0 & 0 \\ 0 & 0 & -i \\ 0 & i & 0 \end{pmatrix} \qquad \lambda_8 = \begin{pmatrix} 1 & 0 & 0 \\ 0 & 1 & 0 \\ 0 & 0 & -2 \end{pmatrix} \frac{1}{\sqrt{3}}.$$

They satisfy the relationships

$$\operatorname{tr}\lambda_c\lambda_{c'} = 2\delta_{cc'}, \qquad \vec{\lambda}\cdot\vec{\lambda} \equiv \sum_c (\lambda_c)^2 = \tfrac{16}{3}I \qquad (1.3)$$

Each of the $(\lambda_c)^2$ is diagonal. The commutators of the λ's are given by

$$[\lambda_c, \lambda_d] = 2\,i\,f_{cde}\lambda^e, \qquad (1.4)$$

where the structure constants f_{cde} are completely antisymmetric in the three indices. The nonvanishing values are given by

$$f_{123} = 1$$
$$f_{147} = f_{246} = f_{257} = f_{345} = -f_{156} = -f_{367} = \tfrac{1}{2}$$
$$f_{458} = f_{678} = \sqrt{3}/2.$$

The upper and lower indices have the same meaning for both the λ's and the f's. g_s is the strong coupling constant. According to renormalization group arguments, it is not a fixed parameter, but rather must be treated as a running coupling constant dependent upon the characteristic length or momentum scale (Q) of the processes involved. The *effective* coupling constant $\alpha_s(Q^2) = g_s^2/4\pi$ is given in perturbation theory by

$$\alpha_s(Q^2) = \frac{12\pi}{(33 - 2n_f)\ln(Q^2/\Lambda_{QCD}^2)}, \qquad (1.5)$$

where n_f is the number of flavors and Λ_{QCD} is the QCD momentum scale; experimentally, $\Lambda_{QCD} \approx$ 150-200 MeV. The formula displays asymptotic freedom, since $\alpha_s \to 0$ as $Q^2 \to \infty$. The formula strongly suggests that the number of flavors is limited to $n_f < 17$, which is comfortably within the current "best bet" of 6.

The antisymmetric gauge (gluon) field tensor is given by

$$F_{\mu\nu}^a = \partial_\mu A_\nu^a - \partial_\nu A_\mu^a + g_s f_{bc}^a A_\mu^b A_\nu^c, \qquad (1.6)$$

where the $f_{abc} = f_{bc}^a$, etc. are the SU(3) structure constants given above. The terms proportional to the $f's$ gives rise to the nonlinear. non-Abelian complications of the theory. From the Lagrangian follow the field equations

$$\partial^\mu F_{\mu\nu}^c = \tfrac{1}{2}g_s\overline{\psi}\gamma_\nu\lambda^c\psi - g_s f_e^{cd} A_d^\mu F_{\mu\nu}^e. \qquad (1.7)$$

The time $(\nu = 0)$ part of this equation is Gauss's law: The first term on the right hand side contains the quark color charge: the second term contains the color charge carried by the gauge fields, and again manifests the nonlinearity of the theory.

1.3 Why Modeling?

Because of the intractability of the QCD equations, and the current limitations of lattice gauge calculations, researchers have been lead to modeling. *Modeling* differs from *approximating* a theory in the following way: In the latter case one begins with what is assumed to be the basic theory and then makes mathematical approximations, which may or may not be justified, in order to render the problem tractable. In modeling, one proffers a physical system which is well-defined, possesses the essential properties of the basic theory, and is simpler to solve, but is generally not equivalent to the basic theory.

QCD modeling has a nice analogy with nuclear modeling. Even when one believed that one had a good theory of nuclear forces, the many body problem at first appeared to be intractable. The shell, optical, and collective models were invented to describe nuclear structure and reactions. When Brueckner theory was introduced, its first goals were to reproduce the empirically derived parameters of the phenomenological models. Similar examples can be found in atomic physics, solid state physics, superconductivity, etc. Of course modeling is also used when there is *no* basic theory.

The goal of modeling here is to bridge the gap between QCD and experiment in the realm of GeV energies and fm distances. At a minimum, the model should contain the following essential features of QCD: Absolute color confinement and asymptotic freedom, along with the elementary properties of the fundamental quark and gluon fields.

1.4 The MIT model

The prototype of bag models is the MIT bag model (Chodos, *et al.*, 1974; see also Bogoliubov, 1968). In its original form, it is elegantly simple. Quarks move freely within a cavity of radius R. The boundary condition on the quark wave function is identical to that obtained by assuming a scalar potential which is zero inside the cavity and infinite outside. There is an energy density B within the volume of the bag; B can also be regarded as the pressure of the vacuum on the bag. Furthermore, there is a chromo-dielectric function which is unity within the bag and zero outside.

Inside the bag, the quarks satisfy the Dirac wave equation

$$(-i\vec{\alpha} \cdot \vec{\nabla} + \beta m)\psi = \epsilon\psi, \qquad (1.8)$$

where m is the flavor (current) mass matrix. The boundary condition on the quark wave function is

$$-i\vec{\gamma}\cdot\hat{n}\,\psi(R) = \psi(R)\,,\qquad\qquad (1.9)$$

where $\hat{n}$ is the normal to the surface. For the case of a spherical bag, the solutions are characterized by the Dirac quantum number κ [see Eq. (3.11)] and for $m = 0$ can be written in the (unnormalized) form

$$\psi_\kappa = \begin{pmatrix} j_{-\kappa-1}(\omega r/R) \\ i\vec{\sigma}\cdot\hat{r}\,j_{-\kappa}(\omega r/R) \end{pmatrix} \mathcal{Y}^l_{jm}(\hat{r}) \qquad \kappa < 0\,. \qquad (1.10\,a)$$

and

$$\psi_\kappa = \begin{pmatrix} j_\kappa(\omega r/R) \\ -i\vec{\sigma}\cdot\hat{r}\,j_{\kappa-1}(\omega r/R) \end{pmatrix} \mathcal{Y}^l_{jm}(\hat{r})\,. \qquad \kappa > 0 \qquad (1.10\,b)$$

Here the $\mathcal{Y}^l_{jm}$ are the spherical harmonic two component spinors [see Eq. (3.10)], and the j_l are the spherical Bessel functions that are regular at the origin. The ω's are the dimensionless eigenvalues determined by the boundary condition, which yields the transcendental equations

$$j_{-\kappa-1}(\omega) = j_{-\kappa}(\omega) \qquad \kappa < 0 \qquad\qquad (1.11\,a)$$

and

$$j_\kappa(\omega) = -j_{\kappa-1}(\omega)\,. \qquad \kappa > 0 \qquad\qquad (1.11\,b)$$

The upper and lower components for the $\kappa = -1$ ($s_{1/2}$) state are displayed in Fig. 1.1.

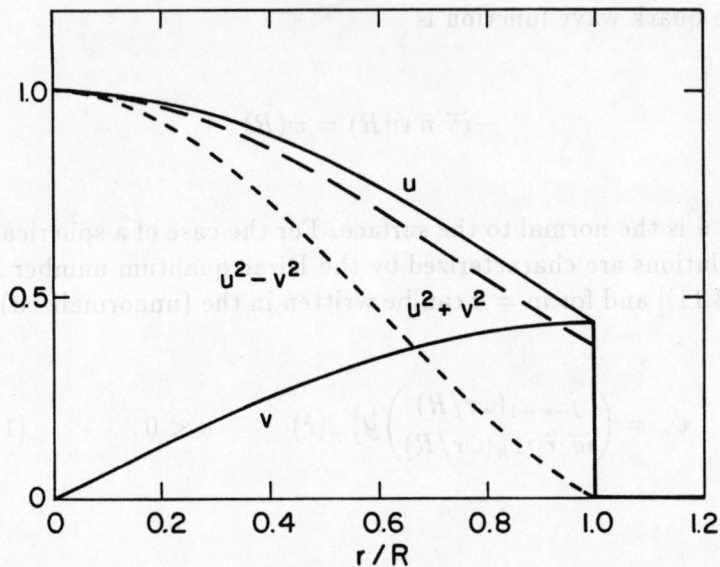

Figure 1.1. Upper (u) and lower (v) components for the lowest positive energy, $\kappa = -1$, state in the MIT bag. The density $u^2 + v^2$ and scalar density $u^2 - v^2$ are also shown. Compare Figs. 3.1 and 8.3.

The first few energy levels for zero mass quarks are

$$\kappa = -1 \ (s_{1/2}): \ \omega = 2.043, \ 5.396, \ 8.578, \ \cdots$$

$$\kappa = +1 \ (p_{1/2}): \ \omega = 3.812, \ 7.002, \ \cdots$$

$$\kappa = -2 \ (p_{3/2}): \ \omega = 3.204, \ 6.758, \ \cdots$$

$$\kappa = +2 \ (d_{3/2}): \ \omega = 5.123, \ 8.408, \ \cdots$$

Note the inverted spin-orbit coupling (with respect to the hydrogen atom): for the same l, the higher j lies lower. Thus the $p_{3/2}$ state lies lower than the $p_{1/2}$. This is characteristic of scalar potentials in contrast to vector potentials.

Adding the volume energy to the quark energy, one obtains for N_q quarks

$$E = \frac{N_q \omega_{-1}}{R} + \tfrac{4}{3}\pi R^3 B. \tag{1.12}$$

The minimum in E occurs at $R = R_0$ where

$$R_0^4 = \frac{N_q \omega_{-1}}{4\pi B}.$$
(1.13)

The bag constant B is the only parameter to this point. One can use (1.12) and (1.13) to obtain various relationships among E, R_0 and B. For example, one can relate the mean baryon (nucleon or delta, $N_q = 3$) mass $\overline{m}$ to the mean radius $\overline{R}$

$$E(\overline{R}) \equiv \overline{m} = \frac{4\,\omega_{-1}}{\overline{R}} = \frac{8.172}{\overline{R}}$$
(1.14)

and

$$B = \frac{3\,\overline{m}^4}{1024\,\pi\omega_{-1}^3} = 1.0936 \times 10^{-4}\overline{m}^4.$$
(1.15)

With $\overline{m} = 1087$ MeV, (1.14) gives $\overline{R} = 1.48$ fm, and (1.15) gives $B = 19.9$ MeV/fm^3. Through first order in α_s, gluon corrections split the baryon masses and radii symmetrically about the mean values. Details of gluonic corrections are given in Chapter 6. For present purposes we can simply add to (1.12) a gluonic correction term proportional to α_s/R. One can then relate the change in bag radius to the change in bag energy by

$$\frac{\delta R}{\overline{R}} = \frac{\pm\frac{1}{2}\Delta m}{3\overline{m}} \qquad \text{for the} \quad \binom{\Delta}{N},$$
(1.16)

where $\Delta m = m_\Delta - m_N = 297$ MeV. This gives a correction to the bag radius of $\pm 4.6\%$, bringing the nucleon bag radius to 1.41 fm.

In the work of DeGrand *et al.* (1975), a term representing the zero-point gluon energy in a cavity, the Casimir (1948) effect, was introduced phenomenologically as $-Z_0/R$. (It was also intended to incorporate other effects.) This is equivalent, for the baryon, to replacing ω_{-1} by $\omega_{-1} - \frac{1}{3}Z_0$ in Eqs. (1.12) - (1.15). Eq. (1.16) is not affected. With the value $Z_0 = 1.84$, which they obtained by fitting data, they found $R_N = 0.99$ fm and $B = 57.5$ MeV/fm^3.

The Casimir effect is both problematical and controversial. It certainly cannot be used without some form of cut-off regularization. This can be seen by noting that the energy of an empty bag,

$$-\frac{Z_0}{R} + \tfrac{4}{3}\pi R^3$$

with $Z_0 > 0$, is monotonic and unbounded from below. Thus a vacuum sprinkled with holes (or even one hole) would be unstable. The cut-off radius for a zero energy bag would be $(3Z_0/4\pi B)^{1/4}$. For the parameters quoted in the previous paragraph, this would yield a (minimum) radius of 1.1 fm, which is inconsistent with the hadron bag radii derived in the calculations. Furthermore, there are ambiguities in the derivation of the term, and even in the sign of the term. (Casimir's parallel capacitor is a conductor, whereas the MIT cavity is surrounded by a perfect dia-electric).

The quark density distribution drops approximately parabolically with radius to a value at the surface equal to 0.3800 of the central value, see Fig. 1.1. The rms quark radius is $0.7290\,R$ compared with $\sqrt{3/5}R = 0.7746\,R$ for a uniform distribution. That is to say, the bag radius is 6% *larger* than what is called the equivalent (charge) radius.

The nucleon magnetic moments are given by $\mu_n = -\tfrac{2}{3}\mu_p$ with $\mu_p = 2.20/2m$, where $1/2m$ is the magneton; the experimental value of μ_p is 2.7928456. The ratio of the axial and vector coupling constants is given in the model by $g_A/g_V = 1.09$ compared with 1.26 for experiment.

1.5 Evolution of the MIT model

Violation of the conservation of axial current (CAC) for systems of massless quarks has plagued bag models from the beginning. (In the presence of massless pions, CAC becomes chiral invariance.) In the MIT model, the axial current is finite in the radial direction at the inner edge of the bag boundary; the divergence of the axial current is infinite across the boundary. In order to remedy this problem, Chodos and Thorn (1975) and Vento, Rho, Nyman, Jun and Brown (1980) coupled an elementary pion field to the bag surface so as to render the axial current continuous. For pions of finite mass, this yielded partial conservation of the axial current (PCAC).

A highly successful program involving quark-pion bags was undertaken by Miller, Thomas and Thèberge (1980) under the name of the cloudy bag model (CBM).

A hybrid model which divides the space of a hadron into an interior MIT bag and an exterior topological Skyrme model has been dubbed the Cheshire cat model (Nadkarni, Nielsen and Zahed, 1985; Vepstas, Jackson and Goldhaber, 1984).

1.6 Why a soliton model?

Static, boundary condition bag models such as the MIT and CBM, have had considerable success in reproducing the spectra and properties of low-lying hadronic states involving light quarks. A difficulty which all such models encounter, however, has been the handling of the *dynamics* of the confinement mechanism (*cf*. Hasenfratz and Kuti, 1978). Soliton models have the important feature that confinement is effected by the intervention of a quantum mechanical scalar field. The effective Lagrangian contains the time derivatives of the fields so that a Hamiltonian can be constructed which contains the field and its conjugate momentum. Methods familiar from nuclear theory can therefore be used to construct fully quantal states of the system. In doing so, one can employ, for example, the coherent state (or, more generally, the single mode) approximation for the scalar field part of the state vector. This is related to the mean field approximation, but is quantal.

[A full treatment of the gluons as dynamical fields still leads to complications. Satisfaction of a gauge condition, for example as in QED, may eliminate a time derivative (Hasenfratz and Kuti, 1978); there are well known techniques for quantization in that case.]

The Lagrangian for the soliton model is the usual QCD Lagrangian supplemented by a non-linear scalar sigma field, which may be interpreted as representing the gluon condensate arising from the non-linear interactions of the color fields. Since the gluons are also represented in the Lagrangian, there is clearly double counting, in principle at least. This does not arise to the order of one gluon exchange, since the sigma field is color-singlet, and one gluon always carries color; two gluon exchange could involve two gluons in a color singlet state and hence could also be contained in the exchange of a sigma quantum. The Lagrangian also contains a color-dielectric function which is a function of the sigma field. The form of the dielectric function assures color confinement.

Parameters associated with the sigma field are adjusted (regularized) to yield physical results for calculated hadronic properties. If we could solve the model Lagrangian exactly, we would obtain the exact QCD results when the model parameters are adjusted to decouple the sigma field from the system. Since the parameters of the sigma field are to be readjusted at every level of the calculation to fit key data, the results of calculations should converge to the exact QCD values. Of course, no one has yet been clever enough to calculate hadronic properties exactly in QCD; so long as the calculations based on the soliton model remain at a relatively simple level, the model must be regarded as phenomenological.

Using a phenomenological form for the gluon propagator, Cahill and Roberts (1985) have proffered an interesting "derivation" of the soliton model.

We will see in Chapter 8 how the model can incorporate CAC (or chiral invariance).

1.7 The variegate world of solitons

Solitary waves were discovered by J. Scott Russell (1845) who observed stable, large amplitude water waves generated by a horse-drawn boat in a narrow channel. The nonlinear hydrodynamic equations admit solutions of finite amplitude waves which propagate with stable wave form, without dispersion. Zabusky and Kruskal (1965) coined the name *soliton* and the term has stuck and has been extended to a variety of nonlinear wave phenomena. In hadronic physics, it is used for structures built of Bosons.

One classification of solitons is according to "topological" and "nontopological" (*cf.* Lee, 1981). Topological solitons have at least two phases of the system which are degenerate in energy density, *i.e.* a degenerate vacuum. Thus the space can be divided into topologically distinct regions of low energy. To transform a region from one state to the other would take finite (or infinite) energy, leading to stability of the topology classically. Nontopological solitons acquire their stability as a result of the presence of some conserved charge, *e.g.*, baryon number for quarks. The Skyrme model (Skyrme, 1961; Witten, 1983) is a famous example of topological solitons.

A variety of models of QCD in the category of nontopological solitons deserve more attention than are reported here. These include the linear sigma model (Birse and Banerjee, 1984, 1985; Kahana, Ripka and Soni, 1984) and the very extensive work of Celenza and Shakin (1987) on a primarily linear version.

Chapter two

THE MODELS

2.1 Origins

Friedberg and Lee (1977, 1978) based their nontopological soliton model on the dielectric properties of the vacuum. The physical vacuum is a complex structure containing virtual excitations of quark pairs and other objects. In particular, it contains strongly correlated gluons in color singlet states—the gluon condensate. They introduced an auxiliary scalar field, σ, which can be identified with the gluon condensate, with the quantum numbers of the vacuum. In the region of the physical vacuum, the σ-field attains a nonzero value, σ_v. There is also a higher, metastable state of the system at $\sigma = 0$. (Of course, a c-number shift in the σ-field could be employed to make the vacuum value of the scalar field zero.) The state $\sigma = 0$ corresponds to the region within which the quarks are nearly free and assume their *current* masses; it is the region where perturbative QCD is assumed to be valid. Friedberg and Lee patterned their soliton model on a nonlinear scalar model introduced earlier by Lee and Wick (1974) to describe a possible high density state of nuclear matter.

The concept of a model of a nonlinear scalar field interacting with quarks is to be found in a number of works during this period. Of particular note is the SLAC bag model (Bardeen, *et al.*, 1975). The SLAC model used a special quartic form for the scalar field self-interaction (see Sec. 2.2) which concentrated the quarks in a shell. However, the appendix in the paper also contains a discussion of the more general quartic form considered by Friedberg and Lee. Other authors who studied quartic forms include Vinciarelli (1972) and Huang and Stump (1976). A model of confinement in a linear scalar field was proposed by Rafelski (1977). The key feature of the Friedberg-Lee model, however, is the dielectric property of the vacuum.

Various attempts have been made to derive the dielectric properties of the medium: Mandelstam (1979) introduced the dual superconductivity model, and this has been pursued most effectively by Baker, Ball and Zachariasen (1986). There is the interesting work of Nielsen and Patkos (1982) as extended by Chanfray, Nachtmann and Pirner (1984). While the dual superconductivity model results are encouraging, it does not represent a rigorous derivation and certain parameters must be inputted phenomenologically. Lattice gauge calculations would be extremely helpful in guiding the model, for example in comparing corresponding flux tube calculations. To date the lattice calculations are not at the degree of precision or reliability to give that guidance.

2.2 The Lagrangian

The non-topological soliton model (Friedberg and Lee, 1977, 1978; Lee, 1979, 1981) is described by a covariant, gauge-invariant Lagrangian density

$$\mathcal{L} = \mathcal{L}_q + \mathcal{L}_\sigma + \mathcal{L}_{q,\sigma} + \mathcal{L}_G, \qquad (2.1)$$

where the various components have the following meaning:

The Dirac term is

$$\mathcal{L}_q = \overline{\psi}(i\gamma^\mu D_\mu - m)\psi, \qquad (2.2)$$

where m is the flavor matrix for current masses. The current quark masses were displayed in Table 1.1. Again, the field operator ψ is taken to incorporate flavor and color as well as the Dirac indices, and D_μ is the gauge-covariant derivative (1.2).

The scalar field term is

$$\mathcal{L}_\sigma = \tfrac{1}{2}\partial_\mu\sigma\,\partial^\mu\sigma - U(\sigma). \qquad (2.3)$$

The fermion-scalar interaction term is

$$\mathcal{L}_{q,\sigma} = -g(\sigma)\overline{\psi}\psi\,. \tag{2.4}$$

This term explicitly breaks chiral invariance since $g(\sigma)$ is an effective mass. Below we discuss various forms for $g(\sigma)$.) In chapter 8 we show how the the model can be solved with this term absent.

The gluonic part of the Lagrangian is

$$\mathcal{L}_G = -\tfrac{1}{4}\kappa(\sigma)F^c_{\mu\nu}F^{\mu\nu}_c\,. \tag{2.5}$$

The gauge field tensor is given by

$$F^a_{\mu\nu} = \partial_\mu A^a_\nu - \partial_\nu A^a_\mu + g_s f^a_{bc}A^b_\mu A^c_\nu\,, \tag{2.6}$$

where the $f_{abc} = f^a_{bc}$, etc. are the SU(3) structure factors given in Sec. 1.1

The sigma-field self interaction is taken to be of the quartic form

$$U(\sigma) = \frac{a}{2}\sigma^2 + \frac{b}{3!}\sigma^3 + \frac{c}{4!}\sigma^4 + B\,, \tag{2.7}$$

see Fig. (2.1). The constants a, b and c are fixed within a range so that $U(\sigma)$ has two minima, one at $\sigma = 0$ and a lower minimum at $\sigma = \sigma_v$, the vacuum value:

$$\sigma_v = \frac{3\,|b|}{2\,c}\left[1 + \sqrt{1 - \tfrac{8}{3}ac/b^2}\right] \tag{2.8}$$

for $b \leq 0$. It is convenient to introduce the "family" parameter $f = b^2/ac$ which characterizes the shape of $U(\sigma)$. The constant B is determined to make $U(\sigma_v) = 0$; then $U(0) = B$ has the same meaning as the MIT bag constant, namely the energy density of deconfinement, or bag pressure. For $f = 3$, we have $B = 0$. The polynomial form for $U(\sigma)$ is limited to fourth order for simplicity and renormalizability, although the latter is not compelling, see chapter 9.

The quantities have the following dimensions of length (L):
$[\sigma] = \mathrm{L}^{-1}$, $\quad [a] = \mathrm{L}^{-2}$, $\quad [b] = \mathrm{L}^{-1}$, $\quad [c] = \mathrm{L}^0$, $\quad [B] = \mathrm{L}^{-4}$.

Nontopological Solitons

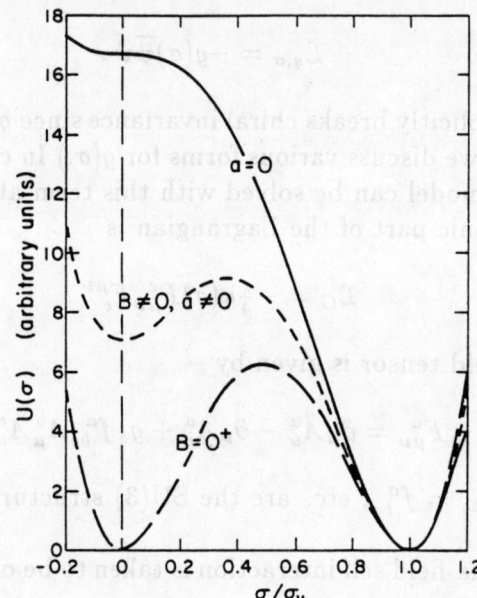

Figure 2.1
Three forms
for $U(\sigma)$.

[For comparison, the SLAC model (in the same notation) is of the form

$$U(\sigma) = \frac{c}{4!}(\sigma^2 - \sigma_v^2)^2$$

which corresponds to $f = 3$. The MIT bag can be obtained by a suitable limit of the parameters (Friedberg and Lee, 1978).]

The dielectric function $\kappa(\sigma)$ satisfies the following conditions in order to guarantee absolute color confinement:

$$\kappa(0) = 1, \qquad \kappa(\sigma_v) = 0. \qquad (2.9\,a,\,b)$$

In addition, the condition

$$\kappa'(\sigma_v) = 0 \qquad (2.9\,c)$$

must be imposed in order to assure regular behavior for σ as $\sigma \to \sigma_v$ (see Sec. 6.3). The particular form of $\kappa(\sigma)$ is not crucial so long as the above properties are satisfied. One particular choice (Bickeböller *et al.*, 1985) has the further properties $\kappa(\sigma < 0) = 1$ and $\kappa'(0) = \kappa''(0) = 0$. Since $\kappa'(\sigma)$ and $\kappa''(\sigma)$ appear in the equation for the gluon propagator, these derivatives were made continuous at $\sigma = 0$. The form used by Bickeböller *et al.* is

$$\kappa(\sigma) = 1 + \theta(\sigma)x^3(3x - 4), \qquad (2.10)$$

with $x = \sigma/\sigma_v$; it can be generalized to

$$\kappa(\sigma) = 1 + \theta(\sigma)x^n(nx - (n + 1)). \qquad (2.11)$$

A restriction to $n > 2$ assures continuity of the second derivative at $\sigma = 0$, but even this is not required for cases where σ does not go through zero, such as are encountered in Chapter 8.

It is important to distinguish statements about the *model* and statements about *approximations* to the model. The model yields absolute color confinement and hence absolute confinement of quarks in color singlet clusters. It is free of the color Van der Waals problem. This is not true of various levels of approximation. These matters are expanded upon below.

Except for the $\kappa(\sigma)$ in (2.5), the model is renormalizable. Since it is already an effective model, it could be argued that there is no need for renormalizability. The sigma field is to be interpreted as a composite field, *e.g.* a gluon condensate. Loop corrections to the sigma field are implicitly included in the effective Lagrangian and it would be double counting to add these in again. As the level of approximation is increased, implicit double counting is handled by readjustment (regularization) of the model parameters.

Note that if we set $g(\sigma) = 0$ and $\kappa = 1$, the sigma field decouples and (2.2) plus (2.5) yield the exact QCD Lagrangian. Since at every level of approximation the model parameters are to be readjusted to yield physical results, the model has the possibility of converging to the exact QCD theory (as more exact solutions are obtained) with the decoupling of the σ field, and the recession of any required cutoff to infinity. This also requires a parameter to let $\kappa \to 1$.

In Chapter 8 we examine the consequences of setting $g(\sigma) = 0$. The Lagrangian is then chirally invariant. Color confinement is effected through gluonic interactions including the quark self energies. An *effective* quark-sigma coupling arises which leads to finite hadronic masses. However, a massless Nambu-Goldstone meson must emerge to restore chiral invariance.

The original Friedberg-Lee (F-L) form for $g(\sigma)$ is

$$g_{FL}(\sigma) = g_0\sigma. \qquad (2.12)$$

Because the coupling is linear in σ, the theory is renormalizable (except for the dielectric function). Without gluonic interactions, however, it does not lead to absolute confinement.

The Nielsen-Patkos (1982) (N-P) model is similar in structure to the Friedberg-Lee model, with identification of a scalar field χ with the F-L σ through

$$\chi = 1 - \sigma/\sigma_v. \qquad (2.13)$$

The scalar field interaction term $U(\chi) = U(\sigma)$ has its minimum at $\chi = 0$, which is the vacuum value of χ. The inside-the-bag value is $\chi \approx 1$. The dielectric function is taken to be $\kappa = \chi^4$. The quark-chi coupling factor in the Lagrangian is

$$g_{NP}(\chi) = \frac{g_0}{\sqrt[4]{\kappa(\chi)}} = \frac{g_0}{\chi}. \qquad (2.14)$$

The scalar confinement potential leads to absolute confinement irrespective of color. The form is non-renormalizable, but this may not be a serious consideration (see Chapter 9). The effective quark mass inside the bag is approximately g_0 plus the current quark mass, or one can subtract g_0 from the coupling term to make the inside mass just the current mass. This model has been pursued by several authors. (Chanfray, Nachtmann and Pirner, 1984; Schuh and Pirner, 1986; Pirner, Wroldsen and Ilgenfritz, 1987; Williams and Thomas, 1986; Dodd, Williams and Thomas, 1987).

Bayer, Forkel and Weise (1986), also building on the F-L model, have proposed coupling of the form

$$g_{BFW}(\sigma) = \frac{g_0\sigma}{\sqrt{\kappa(\sigma)}}, \qquad (2.15)$$

with

$$\sqrt{\kappa(\sigma)} = (1 - \sigma/\sigma_v)^\alpha. \qquad (2.16)$$

Although BFW state the condition that $\alpha > 0$, we see that condition (2.9 c) demands $\alpha > \frac{1}{2}$. The effective quark mass inside the bag is $g_0\sigma$ plus the current quark mass.

As will be seen in Chapter 8, quark self-energy based on chromo-dielectrics for massive quarks leads to an *effective* confinement potential

$$g_{cd}(\sigma) = g_0\left(\frac{1}{\kappa} - 1\right). \qquad (2.17)$$

This also leads to absolute confinement, and has the desirable property that the inside quark mass is very nearly the current quark mass (Fai, Perry and Wilets, 1988). But note that color confinement requires explicit calculation of the chromoelectric self and mutual energies.

2.3. Parameters of the F-L model

The five model parameters are:

    a, b, c. which are contained in $U(\sigma)$.

    g_0, the σ-quark coupling constant in the various forms discussed. Later we will see that in the chromodielectric model (CDM), g_0 is related to the strong coupling constant α_s and a QCD cutoff parameter.

    $\alpha_s = g_s^2/4\pi$, the strong coupling constant.

    There are also the possible forms of $\kappa(\sigma)$.

Some of the key data to be fit are:

(1) The mean baryon mass $\overline{m} = \frac{1}{2}(m_N + m_\Delta) = 1087$ MeV;

(2) The proton root-mean-square charge radius $<r_p^2>^{\frac{1}{2}} = 0.83$ fm;

(3) The nucleon magnetic moments,
$\mu_p = 2.7928456$ nm and $\mu_n = -1.9130418$ nm, where a nuclear magneton is $e\hbar/2m_p c \equiv 1/2m_p$;

(4) The ratio of the axial to vector coupling constants $g_A/g_V \simeq 1.26$;

(5) The light meson masses, $m_\pi = 138$ MeV, $m_\rho \approx m_\omega \approx 780$ MeV;

(6) The delta-nucleon mass splitting $m_\Delta - m_N = 297$ MeV;

(7) The coefficient of the linear term in the heavy $q\bar{q}$ potential (from charmonium, bottomonium, *etc.*), called the string tension or flux tube potential, $\theta \approx 925$ MeV/fm (Eichten, *et al.*, 1980; Buchmüller, 1982; Gupta *et al.*, 1982; Moxhay and Rosner, 1983); it is sometimes denoted by $1/a^2$.

There are also very interesting quantities which have not yet been measured. We look to both experiment and lattice gauge calculations to determine their values. These include:

(9) The bag constant B, which is the volume energy of the cavity. In the simple MIT model it has a value of about 20 MeV/fm³; with gluonic interactions and the $-Z_0/R$ term it is fitted to 57 MeV/fm³. It is also related to the deconfinement transition temperature, although the connection is clearly model-dependent.

(10) The glueball mass. We identify the 0^{++} glueball with excitations of the sigma field according to $m_{GB}^2(0^{++}) = U''(\sigma_v)$.

Of course, the goal of a model is to fit as much data as possible. The above data are selected to be "key" in the sense that the five model parameters are to be adjusted at every level of approximation to obtain good fits to these quantities. Other experimental data are then "predicted" by the model.

Rosina, Schuh and Pirner (1986) have presented relationships among various of the F-L parameters, including temperature.

The program, then, is to obtain accurate numerical solutions to the model. An appropriate starting point for light quarks ($m_q \ll m_{GB}$) is the mean field approximation, while for massive quarks ($m_q \gg m_{GB}$) it is the potential obtained in the adiabatic (Born-Oppenheimer) approximation.

Chapter three

THE MEAN FIELD APPROXIMATION: CLASSICAL AND QUANTAL

3.1 Field equations without gluons

As a preliminary to dynamical calculations, we consider first static solutions to the field equations. The simplest of these is the mean field approximation (MFA). We first neglect gluonic interactions and restrict consideration to valence quarks (no vacuum distortion at this stage) and use $g_{FL}(\sigma)$. This approximation does not lead to absolute confinement of the quarks, but is useful for the low-lying quark states. (Absolute confinement will be effected in Chapters 6 and 8.) The Hamiltonian can then be written

$$H = \int d^3r \, \mathcal{H}(\vec{r}),$$

$$
\begin{aligned}
\mathcal{H}(\vec{r}) = &\, \psi^\dagger(\vec{r}) \left[-i\,\vec{\alpha}\cdot\vec{\nabla} + \beta g(\sigma(\vec{r})) \right] \psi(\vec{r}) \\
&+ \tfrac{1}{2}\pi(\vec{r})^2 + \tfrac{1}{2}|\vec{\nabla}\sigma(\vec{r})|^2 + U(\sigma),
\end{aligned}
\tag{3.1}
$$

where $\pi = \dot{\sigma}$. We can always separate the sigma field according to

$$\sigma = \sigma_0(\vec{r}) + \sigma_1, \quad \pi = \pi_0(\vec{r}) + \pi_1, \tag{3.2}$$

where $\sigma_0(\vec{r})$ is a time-independent c-number and σ_1 is the quantum fluctuation operator. Because σ_0 is chosen to be static, it follows that $\pi_0 = 0$. The operators satisfy the usual equal-time Bose commutation relations,

$$[\pi(\vec{r}, t), \sigma(\vec{r}\,', t)] = [\pi_1(\vec{r}, t), \sigma_1(\vec{r}\,', t)] = -i\delta^3(\vec{r} - \vec{r}\,'). \tag{3.3}$$

Similarly, we can represent the quark field operators by

$$\psi = \sum_k c_k \psi_k(\vec{r}), \tag{3.4}$$

where the c_k satisfy the equal-time Fermi anti-commutation relations

$$[c_k(t), c_{k'}^\dagger(t)]_+ = \delta_{kk'}, \tag{3.5}$$

and the $\psi_k(\vec{r})$ are any complete and orthonormal set of spinor-color-flavor functions. Throughout the book, we will work in the Schrödinger picture so that operators are time-independent.

We consider a fixed occupation number of valence quarks (3 quarks for nucleons, a quark-antiquark pair for mesons). The Hamiltonian density can be rewritten

$$\mathcal{H} = \mathcal{H}(\sigma_0) + \mathcal{H}'(\sigma_0)\sigma_1 + \tfrac{1}{2}\mathcal{H}''(\sigma_0)\sigma_1^2 + \tfrac{1}{6}\mathcal{H}'''(\sigma_0)\sigma_1^3 + \tfrac{1}{24}\mathcal{H}''''\sigma_1^4. \tag{3.6}$$

Extremization of the expectation value of H with respect to $\psi_k^\dagger$ (subject to the normalization constraint) and with respect to σ_0, neglecting terms of order σ^2 and higher, leads to the coupled set of equations

$$\left[\vec{\alpha} \cdot \vec{p} + \beta g(\sigma_0)\right]\psi_k = \epsilon_k \psi_k, \tag{3.7}$$

$$-\nabla^2 \sigma_0 + U'(\sigma_0) + g'(\sigma_0) \sum_{k(valence)} \psi_k^\dagger \beta \psi_k = 0. \tag{3.8}$$

The first is a linear eigenvalue problem (if σ_0 is given); the second is a non-linear inhomogeneous equation (if the ψ_k are given). Satisfaction of the extremization conditions also eliminates terms proportional to σ_1 in the expectation value of the energy. In addition to employing the mean field approximation, the set of equations (3.7-8) neglects the negative energy sea of quarks, *i.e.* vacuum distortion. We will return to this matter in Chapter 9.

In the special case that $\sigma_0(\vec{r})$ is spherically symmetric, the Dirac equation functions can be written

$$\psi_k = \begin{pmatrix} u_k(r) \\ i\vec{\sigma} \cdot \hat{r}\, v_k(r) \end{pmatrix} \mathcal{Y}_{\kappa m}, \tag{3.9}$$

where $\mathcal{Y}_{\kappa m} \equiv \mathcal{Y}_{jm}^{\ell}$ is the two-component Pauli spinor harmonic,

$$\mathcal{Y}_{jm}^{\ell} = \sum_{m_{\ell}m_s} <\ell\, m_{\ell}\, \tfrac{1}{2}\, m_s | jm> Y_{\ell m_{\ell}}\, \chi_{m_s}\, . \qquad (3.10)$$

The Dirac quantum number

$$\kappa = \left(j + \tfrac{1}{2}\right)\left(-1\right)^{j-\ell+\frac{1}{2}} \qquad (3.11)$$

denotes both j and ℓ. In this case, Eqs. (3.8) can be replaced by

$$\left(\frac{d}{dr} + \frac{(\kappa+1)}{r}\right) u_k + (g(\sigma_0) + \epsilon_k)\, v_k = 0, \qquad (3.12\,a)$$

$$\left(\frac{d}{dr} - \frac{(\kappa-1)}{r}\right) v_k + (g(\sigma_0) - \epsilon_k)\, u_k = 0, \qquad (3.12\,b)$$

$$-\frac{1}{r}\frac{d^2}{dr^2}\, r\sigma_0 + U'(\sigma_0) + \frac{g'(\sigma_0)}{4\pi} \sum_{k(valence)} (u_k^2 - v_k^2) = 0\,, \qquad (3.12\,c)$$

with the normalization condition (which differs by a factor of 4π from that of Goldflam and Wilets (1982))

$$\int_0^{\infty} r^2 dr \left(u_k^2 + v_k^2\right) = 1. \qquad (3.12\,d)$$

We have assumed here that the quark distribution is also spherically symmetric. Various techniques are available for solving Eqs. (3.12) (*cf.* Horn, Goldflam and Wilets, 1986) or their non-spherical generalizations (Schuh, 1985). Numerical techniques are described in the Appendix.

3.2 Baryon bag states

An example of a solution for the baryon, $(s_{1/2})^3$ is shown in Fig. 3.1.

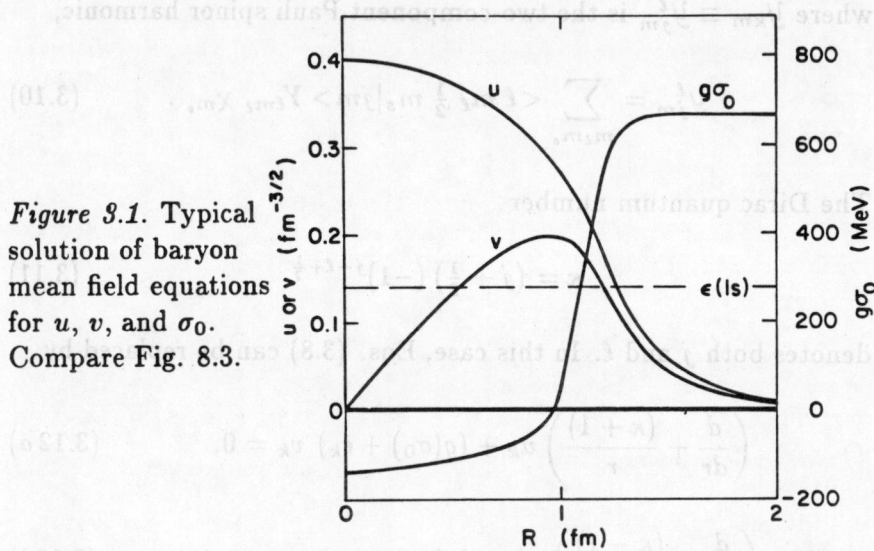

Figure 3.1. Typical solution of baryon mean field equations for u, v, and σ_0. Compare Fig. 8.3.

From the lowest energy $s_{1/2}$ wave functions, nucleon and delta functions can be constructed. Because we are dealing with three quarks, this can be done most simply as follows (*cf.* Lichtenberg 1978):

In color coordinates, the wave function must be a singlet, hence antisymmetric in all pairs; the color function $X_0^C(123)$ is a common factor. The remaining factor must be symmetric in the space, spin and isospin coordinates. Since the three spatial states are identical, it is automatically symmetric in space, so $\Psi(123)$ is also a common factor. Let us write the spatial part as a product of two component spinors in "ρ-space"

$$\Psi(123) = \begin{pmatrix} u \\ i\vec{\sigma}\cdot\hat{r}\,v \end{pmatrix}_1 \otimes \begin{pmatrix} u \\ i\vec{\sigma}\cdot\hat{r}\,v \end{pmatrix}_2 \otimes \begin{pmatrix} u \\ i\vec{\sigma}\cdot\hat{r}\,v \end{pmatrix}_3, \quad (3.13)$$

where here $u = u(r)$ and $v = v(r)$ are understood to be the $s_{1/2}$ upper (large) and lower (small) components of the corresponding Dirac functions.

The proton contains two up (u) quarks and one down (d) quark. We write the isospin factor

$$u(1)u(2)d(3).$$

The uu pair is symmetric in isospin, hence it must be symmetric in spin, namely an $S = 1$ state. The d quark then couples with the pair to a spin of $1/2$. The function thus obtained is not manifestly symmetric in spin-isospin, but can be symmetrized, S, "by hand." We obtain for the proton

$$|p, m> \equiv |\tfrac{1}{2}\tfrac{1}{2}; \tfrac{1}{2}m> = X_0^C(123)\Psi(123)\; S\left\{u(1)u(2)d(3)\times\right.$$

$$\left.\sum_{m_1 m_2 m_3, M} \chi_{m_1}(1)\chi_{m_2}(2)\chi_{m_3}(3) <\tfrac{1}{2}m_1\tfrac{1}{2}m_2|1M><1M\tfrac{1}{2}m_3|\tfrac{1}{2}m>\right\}$$

$$(3.14)$$

in the notation $|I\,m_I; J\,m_J>$, where $S = \tfrac{1}{3!}\sum_P P$ is the symmetrizing operator. In particular,

$$|p\uparrow> \equiv |\tfrac{1}{2}\tfrac{1}{2}; \tfrac{1}{2}\tfrac{1}{2}> =$$

$$X_0^C(123)\Psi(123)\; S\left\{u(1)u(2)d(3)\left[-\sqrt{\tfrac{1}{6}}(\uparrow\downarrow + \downarrow\uparrow)\uparrow + \sqrt{\tfrac{2}{3}}\uparrow\uparrow\downarrow\right]\right\}.$$

$$(3.15)$$

The arrows $\uparrow$ and $\downarrow$ denote spin $+\tfrac{1}{2}$ and $-\tfrac{1}{2}$ respectively. The neutron is obtained by interchanging u and d.

The "stretched" delta state is given trivially by

$$|\Delta^{++}, \tfrac{3}{2}> = X_0^C(123)\Psi(123)u(1)u(2)u(3)\uparrow\uparrow\uparrow. \qquad (3.16)$$

The general Δ state can be written in the product form

$$|\Delta, m_J> \equiv |\tfrac{3}{2}m_I; \tfrac{3}{2}m_J> = X_0^C\Psi X_{3/2, m_I}^I X_{3/2, m_J}^S. \qquad (3.17)$$

Here the X^C, X^I and X^S denote color, isospin and spin eigenstates, respectively, for the three quarks. Note that the color function is antisymmetric and that for the Δ each of the other factors is individually symmetric in the exchange of any two coordinates.

For the nucleon (for example), with 3 quarks in the lowest positive energy $s_{1/2}$ state, we have the following MFA expressions for selected properties: The bag energy or mass is

$$m = <H> = 3\epsilon + \int_0^\infty dr \, r^2 \left[U(\sigma_0) + \frac{1}{2} \left(\frac{d\sigma_0}{dr} \right)^2 \right].$$

(3.18)

The proton mean-square (charge) radius is

$$<r^2> = \int_0^\infty dr \, r^4 \left(u^2 + v^2 \right).$$

(3.19)

The neutron charge distribution is zero at this level of approximation. The proton magnetic moment is

$$\mu_p = <p\uparrow | \hat{z} \cdot \vec{r} \times \vec{\alpha} | p\uparrow> = \frac{2}{3} \int_0^\infty dr \, r^3 uv.$$

(3.20)

(As previously noted, for the MIT model, one finds $\mu_p = 2.20$ nm, and for the SLAC bag 2.65 nm, where a nuclear magneton, nm, is $1/2m$.) The neutron magnetic moment is

$$\mu_n = -\frac{2}{3} \mu_p.$$

(3.21)

This well-known QCD result is better satisfied experimentally than the pion field theory result which gives for the anomalous magnetic moments $\kappa_n = -\kappa_p$.

The isovector-vector current operator is

$$\underset{\rightarrow}{V}{}^\mu = \bar{\psi} \, \underset{\rightarrow}{\tau} \, \gamma^\mu \psi ;$$

(3.22)

the axial current operator is

$$\underset{\rightarrow}{A}{}^\mu = \bar{\psi} \, \underset{\rightarrow}{\tau} \, \gamma^\mu \gamma_5 \psi .$$

(3.23)

The "underarrow" is introduced here to denote a vector in isospin space. The ratio of the axial to vector coupling constants can be expressed by

$$\frac{g_A}{g_V} = <p\uparrow | \tau_3 \gamma_0 \gamma^3 \gamma_5 | p\uparrow> = <p\uparrow | \tau_3 \Sigma_3 | p\uparrow> .$$

(3.24)

We have here used the notation $\vec{\Sigma}$ for the 4×4 Dirac spin matrix. In the present approximation this yields

$$\frac{g_A}{g_V} = \tfrac{5}{3} \int_0^\infty dr\, r^2 \left(u^2 - \tfrac{1}{3}v^2\right) . \tag{3.25}$$

(For the MIT model, one finds $g_A/g_V = 1.09$ and for the SLAC bag 0.57).

The presence of a small, nonvanishing current quark mass leads to a contribution to the baryon mass of

$$\delta m = \sum <\beta> m_q .$$

For the MIT model, $<\beta> = 0.4795$; in the soliton model, it is characteristically 0.5.

3.3 Small amplitude oscillations

In terms of the quark states obtained by solving Eqs. (3.7-8) or 3.12), and the Hamiltonian density (3.6), the Hamiltonian can be written

$$H = E_0 + \sum_k {}' \epsilon_k\, c_k^\dagger c_k + \int d^3r \Big\{ \tfrac{1}{2} \big[\pi_1^2 + |\vec{\nabla}\sigma_1|^2 + U''(\sigma_0(\vec{r}))\sigma_1^2\big]$$

$$+ \tfrac{1}{6} U'''(\sigma_0(\vec{r}))\sigma_1^3 + \tfrac{1}{24} c\sigma_1^4 + g \sum_{k\ell} {}' \overline{\psi}_k(\vec{r})\sigma_1\psi_\ell(\vec{r}) c_k^\dagger c_\ell \Big\}, \tag{3.26}$$

where $E_0 = <\int d^3r\, \mathcal{H}(\sigma_0(\vec{r}))>$. The primes on the sums denotes omission of either single states k or pairs $\{k, l\}$ occupied in the ground state. Note the vanishing of terms linear in σ_1 in the ground state. The quantum part of the soliton field can be expanded in terms of any orthonormal set $\{s_n(\vec{r})\}$:

$$\sigma_1 = \sum_n \left(\frac{1}{2\omega_n}\right)^{1/2} \left(a_n^\dagger s_n^* + a_n s_n\right), \qquad (3.27a)$$

$$\pi_1 = i \sum_n \left(\frac{\omega_n}{2}\right)^{1/2} \left(a_n^\dagger s_n^* - a_n s_n\right). \qquad (3.27b)$$

The commutation relations (3.3) are satisfied independently of the choice of the ω_n. However the Hamiltonian (3.26) simplifies if we *choose* the $s_n(\vec{r})$ and ω_n to be solutions of

$$\left(-\nabla^2 + U''(\sigma_0) - \omega_n^2\right) s_n(\vec{r}) = 0. \qquad (3.28)$$

Then

$$H = E_0 + {\sum_k}' \epsilon_k\, c_k^\dagger c_k + \sum_n \omega_n \left(a_n^\dagger a_n + \tfrac{1}{2}\right) + H_1 + H_3 + H_4. \qquad (3.29)$$

The other parts of the Hamiltonian are given by

$$H_1 = g {\sum_{k\ell n}}' \int \overline{\psi}_k\, s_n\, \psi_\ell d^3 r \left(\frac{1}{2\omega_n}\right)^{1/2} c_k^\dagger c_\ell a_n + h.c., \quad (3.30\,a)$$

$$H_3 = \tfrac{1}{6} \int (b + c\sigma_0)\sigma_1^3\, d^3 r \qquad (3.30\,b)$$

$$H_4 = \tfrac{1}{24} c \int \sigma_1^4\, d^3 r, \qquad (3.30\,c)$$

where σ_1 in H_3 and H_4 is represented in terms of a_n and a_n^+ by (3.27 a). The diagrammatic meaning of these terms is shown in Fig. 3.2, where the solid lines are quark propagators and the wavy lines are σ_1 propagators.

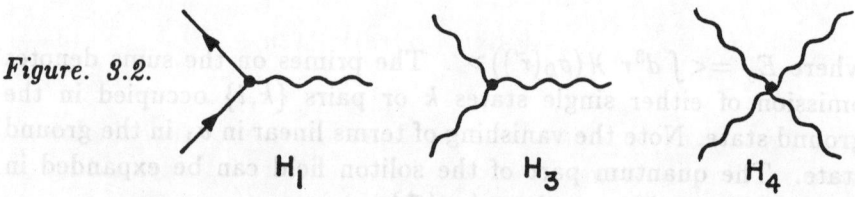

Figure. 3.2.

$$H_I \qquad\qquad H_3 \qquad\qquad H_4$$

Eq. (3.28) defines normal modes for oscillations about the mean field solution. Since we have here considered $\sigma_0(\vec{r})$ to be spherically symmetric, we can set

$$s_n(\vec{r}) \equiv s_{\ell m n}(\vec{r}) = r^{-1} \, u_{\ell n}(r) \, Y_{\ell m}(\theta, \phi) \,, \qquad (3.31)$$

whence (3.28) becomes

$$\left(-\frac{d^2}{dr^2} + \frac{\ell(\ell+1)}{r^2} + U''(\sigma_0(\vec{r})) - \omega_{\ell n}^2 \right) u_{\ell n} = 0 \,. \qquad (3.32)$$

$U''(\sigma_0(r))$ characteristically has a sharp dip in the vicinity of the bag surface, as can be seen in Fig. 3.3. A rough estimate of the eigenfrequencies can be obtained by setting

$$\frac{\ell(\ell+1)}{r^2} + U'' \approx \frac{\ell(\ell+1)}{r_0^2} + U''(\sigma_0(\vec{r}_0)) + (r - r_0)^2 \, \Omega^2 \,, \qquad (3.33)$$

where $r_0 = r_0(\ell)$ is the location of the minimum of the function on the LHS of (3.33) and

$$\Omega^2 = \frac{1}{2} \left[6 \frac{\ell(\ell+1)}{r_0^4} + \frac{d^2}{dr^2} U''(\sigma_0(r)) \right]_{r=r_0} \,. \qquad (3.34)$$

Then

$$\omega_{\ell n}^2 \approx \frac{\ell(\ell+1)}{r_0^2} + U''(\sigma_0(r_0)) + (2n+1)\Omega \,. \qquad (3.35)$$

Note that ω and Ω do not have the same dimensions. For the lowest state we find ω_{00} ranges between 300 and 600 MeV, depending on the parameter set. Note that a problem could arise if $\omega_{\ell n}^2 < 0$.

Figure 3.3.
Schematic
representation of
$U''(\sigma_0)$ showing
surface dip.
(A parameter set
with $B = 0$.)

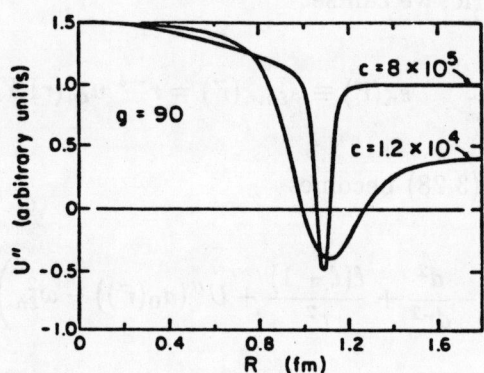

Characteristic quark (ϵ) and soliton (ω) spectra are displayed in Fig. 3.4.

Because of the sharp dip in U'' near the surface, the lowest normal modes are surface modes (*cf.* Iwasaki and Kondo, 1987). The soliton-quark interaction term, H_1, couples the σ_1 excitations to quark particle-hole pairs. $q\bar{q}$ virtual excitations are interpreted as giving rise to a "meson" cloud surrounding the bag or, more specifically, the nucleon. We will return to the meson in Chapter 7.

Through the coupling terms, (3.30), physical states are linear combinations of quark and soliton excitations. This is how Broniowski, Cohen and Banerjee (1987) described the Roper resonance in the context of a color dielectric model.

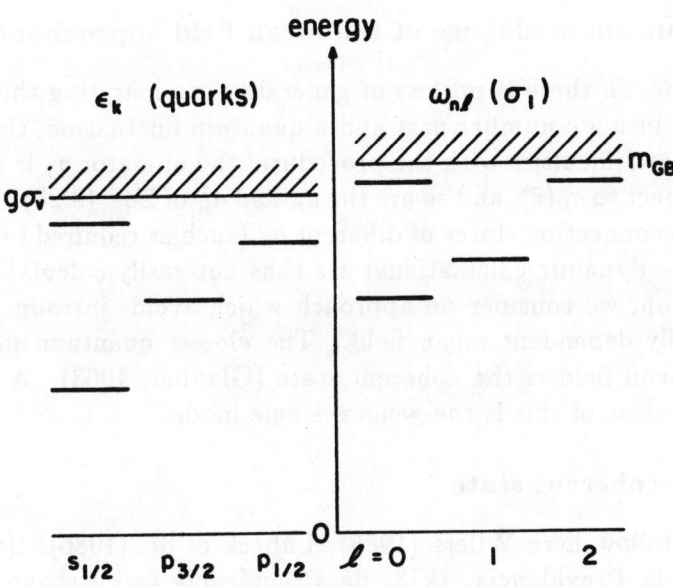

Figure 3.4. Schematic representation of the quark and soliton spectra. The continuum in the quark spectrum is an artifact of the approximation, see Chapter 5.

Although we began with a covariant Lagrangian, the MFA destroys covariance by the selection of a preferred frame. Inclusion of the σ_1 part of the soliton field can restore covariance. For example, the localized MFA bag has $<\vec{P}^2> \neq 0$. A state with $\vec{P} = 0$ is necessarily spread out over all space. Thus the calculated $<r^2>$ increases (ultimately to ∞) as the approximations are improved, and we are alerted to the fact that $<r^2>$ is not a measure of nucleon size. We return to this in Chapter 4.

The terms H_3 and H_4 involve only σ_1 (not quark) operators and lead to a restructuring of the soliton spectra. For present purposes, we will neglect these terms and assume that their effects can be absorbed into the effective parameters, although there may be important physical effects lurking in these terms.

3.4 Quantum analogues of the mean field approximation

Although there is no loss of generality in separating the sigma operator into a c-number part and a quantum fluctuation, there are sometimes difficulties with the procedure: the operator σ_1 is defined with respect to $\sigma_0(\vec{r})$, and so are the a_n and $a_n^\dagger$ of Eqs. (3.27). Matrix elements connecting states of different σ_0 (such as required for recoil and other dynamic calculations) are thus not easily calculable. For this reason, we consider an approach which avoids introduction of a spatially-dependent mean field. The closest quantum analogue of the mean field is the coherent state (Glauber, 1963). A simple generalization of this is the general single mode.

3.5 The coherent state

We follow here Wilets (1985), Lübeck *et al.* (1986). (See especially da Providencia, 1973; da Providencia and Urbano, 1978; Bolsterli, 1979.) We expand σ and π in some complete set of orthonormal functions,

$$\sigma(\vec{r}) = \sigma_v + \sum_n \left(\frac{1}{2\omega_n}\right)^{1/2} \left(a_n^\dagger s_n^*(\vec{r}) + a_n s_n(\vec{r})\right) ,$$

$$\pi(\vec{r}) = i \sum_n \left(\frac{\omega_n}{2}\right)^{1/2} \left(a_n^\dagger s_n^*(\vec{r}) - a_n s_n(\vec{r})\right) , \qquad (3.36)$$

where the ω_n are as yet undetermined. The $\{s_n(\vec{r})\}$ may, for example, be chosen to be plane waves,

$$s_n(\vec{r}) \to s_{\vec{k}}(\vec{r}) = V^{-1/2} e^{i\vec{k}\cdot\vec{r}},$$

with V the box volume. $\sigma(\vec{r},t)$ and $\pi(\vec{r},t)$ satisfy the usual equal time commutation relations independently of the choice of the ω_n if

$$\left[a_n(t),\, a_{n'}^\dagger(t) \right] = \delta_{nn'} ,$$

$$[a_n(t),\, a_{n'}(t)] = \left[a_n^\dagger(t),\, a_{n'}^\dagger(t) \right] = 0 . \qquad (3.37)$$

A "coherent state" (cs) in one mode, say $n = 0$, is obtained by the construction

$$|\lambda> \equiv e^{\lambda a_0^\dagger}|0> , \qquad (3.38)$$

where $a_n|0>= 0$, for all n.

In a temporary, but self-evident, change of notation, we use the relations

$$a_0^\dagger|n> = \sqrt{n+1}\,|n+1>, \qquad a_0|n>= \sqrt{n}\,|n-1>$$

to obtain

$$e^{\lambda a_0^\dagger}|0>= \sum_{n=0}^{\infty} \frac{(\lambda a_0^\dagger)^n}{n!}|0>= \sum_{n=0}^{\infty} \lambda^n \frac{1}{\sqrt{n!}}|n> .$$

Then the coherent state can be shown to be an eigenstate of the annihilation operator a_0,

$$a_0|\lambda>= \sum_{n=1}^{\infty} \lambda^n \frac{1}{\sqrt{(n-1)!}}|n-1>= \lambda|\lambda> . \qquad (3.39)$$

It can be easily verified that

$$<\lambda|\lambda'>= e^{\lambda^* \lambda'} . \qquad (3.40\,a)$$

and that for any operator $O(a_0^\dagger, a_0)$, it follows that

$$<\lambda| : O(a_0^\dagger, a_0) : |\lambda'>= e^{\lambda^* \lambda'} O(\lambda^*, \lambda') . \qquad (3.40\,b)$$

Multiple-mode coherent state vectors can be constructed by taking a product of exponentiated operators:

$$|\lambda_1\lambda_2\cdots> \equiv \prod_n e^{\lambda_n a_n^\dagger}|0>= e^{\sum_n \lambda_n a_n^\dagger}|0> \equiv e^{\Lambda A_0^\dagger}|0> . \qquad (3.41)$$

Again one has

$$a_m|\lambda_1\lambda_2\cdots> = \lambda_m|\lambda_1\lambda_2\cdots> . \tag{3.42}$$

However, (3.41) can be regarded as a single mode state by identifying

$$A_0^\dagger \equiv \Lambda^{-1}\sum_n \lambda_n a_n^\dagger \tag{3.43}$$

to be one member of a transformed set of operators denoted by $A_k^\dagger$, where for normalization

$$\Lambda^2 = \sum_n |\lambda_n|^2 . \tag{3.44}$$

Consider now the normalized coherent state

$$|f; cs> = e^{\sum_n \sqrt{\omega_n/2}\, f_n a_n^\dagger}|0> e^{-\frac{1}{2}\sum_n \omega_n |f_n|^2} \tag{3.45}$$

where the factor $(\omega_n/2)^{1/2}$ is chosen for convenience and f stands for the set of Fourier coefficients of $\sigma_0(r)$, namely $\{f_n\}$. Then

$$<f; cs|\sigma|f; cs> = \sigma_v + \frac{1}{2}\sum_n [f_n s_n(\vec{r}) + f_n^* s_n^*(\vec{r})] \equiv \sigma_0(\vec{r})$$

$$<f; cs| : \sigma^n : |f; cs> = <f; cs|\sigma|f; cs>^n = \sigma_0^n(\vec{r})$$

$$<f; cs|\pi|f; cs> = -\frac{i}{2}\sum_n \omega_n [f_n s_n(\vec{r}) - f_n^* s_n^*(\vec{r})] \equiv \pi_0(\vec{r})$$

$$<f; cs| : \pi^n : |f; cs> = <f; cs|\pi|f; cs>^n = \pi_0^n(\vec{r}) . \tag{3.46}$$

Equations (3.46) look classical, but that is because of the normal ordering operation. In fact, there are quantum fluctuations, and, for example,

$$<f; cs|\sigma^2|f; cs> \neq <f; cs|\sigma|f; cs>^2, \tag{3.47}$$

since here σ^2 is not normal ordered.

Without approximation, the Hamiltonian can be reorganized so that it is expressed in terms of normal-ordered operators only. If we choose the $s_n(\vec{r})$ to be plane waves, say $e^{i\vec{k}_n \cdot \vec{r}}/L^{3/2}$ then with suitable counter terms the new $H \equiv :H:$ is of the form (3.26) except that all operators are normal ordered. The normal ordering does depend upon the choice of ω_n. *In what follows we will take H to be normal ordered with respect to a chosen set of ω_n.*

In the spirit of the MFA, we write the total state vector

$$|\Psi> = \prod_{k(valence)} c_k^\dagger \, e^{\Sigma_n \sqrt{\omega_n/2} f_n a_n^\dagger} |0> e^{-\frac{1}{2}\Sigma_n \omega_n |f_n|^2} . \qquad (3.48)$$

For the static bag case, one may take $\sum_n \omega_n f_n s_n$ to be real, whence $< f; \text{cs}|\pi|f; \text{cs} >= 0$. Then

$$< \Psi|H|\Psi > = H(\sigma_0) , \qquad (3.49)$$

where $H(\sigma_0)$ is given by just Eq. (3.1) with σ replaced by $\sigma_0(\vec{r})$, a c-number. We are again led to the MFA equations. However, the state vector contains fluctuations of the σ-field, and this has consequences for recoil and center-of-mass corrections. Note that in this approximation the mode frequencies ω_n do not even appear. That is because we are still dealing with a static calculation; they will appear in dynamic calculations.

3.6 The general single mode approximation.

One generalization of the coherent state is the general single mode (sm) approximation (Wilets, *et al.*, 1984; Lübeck *et al.*, 1986), which is essentially the Tamm-Dancoff or Tomanaga approximation. Rather than the exponential functional form for the creation operators given in (3.45), one may assume a general function $\mathcal{F}$ of a single mode operator [cf. (3.43)]

$$|f; \text{sm} >= \mathcal{F}\left(\Lambda^{-1} \sum_n \sqrt{\omega_n/2} \, f_n a_n^\dagger \right) |0> . \qquad (3.50)$$

This has been studied by Lübeck *et al.* where the function $\mathcal{F}$ was expanded in a power series:

$$\mathcal{F}(A^\dagger) = \sum_m \frac{F_m}{m!} (A^\dagger)^m . \qquad (3.51)$$

Here $A^\dagger$ is the normalized argument of $\mathcal{F}$ appearing in (3.50). For the cases tested, there were some differences from the coherent state results, but the F_m varied slowly over a fairly large range of m-values and the series converged reasonably rapidly. If the F_m were independent of m, $\mathcal{F}$ would be an exponential and we would have the usual coherent state. The algebra for evaluating matrix elements is somewhat more complicated than for the coherent state, but is still quite tractable. In particular, it follows from the commutation relation $[A, A^\dagger] = 1$ that

$$< 0 | A^m (A^\dagger)^n | 0 > = \delta_{mn}\, n! \, , \tag{3.52}$$

which shows that this expansion corresponds to an orthogonal basis. A normalized state is thus $(n!)^{-1/2}(A^\dagger)^n |0>$. Similar use of the commutation rules yields

$$< 0 | A^m\, a_k\, (A^\dagger)^n | 0 > = \sqrt{\frac{\omega_k}{2}}\, f_k\, \delta_{m,n-1}\, n! \, . \tag{3.53}$$

With this one can evaluate, for example, the expectation value of $\sigma(\vec{r})$ with respect to the coherent state vector (3.50):

$$< f; \mathrm{sm} |\sigma(\vec{r})| f; \mathrm{sm} > = \sigma_v + \frac{\displaystyle\sum_n \frac{1}{n!} F_n\, F_{n-1}}{\displaystyle\sum_n \frac{1}{n!} F_n^2}\, \sum_k f_k s_k(\vec{r}) \equiv \sigma_0(\vec{r}). \tag{3.54}$$

From similar straight-forward, but tedious, use of the commutation rules, one can get expressions for the expectation values of normal-ordered products of the field operators which occur in the Hamiltonian. Variation of the energy with respect to the coefficients F_n leads to a set of linear equations of the form,

$$\sum_m H_{nm}\, F_m = E\, F_n, \tag{3.55}$$

where

$$H_{nm} = \frac{< 3q | A^n H (A^\dagger)^m | 3q >}{(n!m!)^{1/2}} \, . \tag{3.56}$$

Here $|3q>$ is a state vector with three valence quarks and no σ excitation. The solution to this set of equations with the lowest eigenvalue is the nucleon ground state in this approximation.

Results of calculations employing single mode and coherent state approximations are presented in Chapter 7.

Chapter four

PROJECTION AND BOOST

4.1 The center of mass problem

A composite structure localized in a particular reference frame is a wave packet containing a distribution of momentum components. This is the case for a bag described in the MFA. The total linear momentum operator is, neglecting gluons,

$$\vec{P} = -\tfrac{1}{2} \int d^3r \left[i\psi^\dagger \overset{\leftrightarrow}{\nabla} \psi + \{\pi, \vec{\nabla}\sigma\} \right], \tag{4.1}$$

where $\overset{\leftrightarrow}{\nabla}$ means the difference of terms obtained by operating with the nabla to the left and right (the hermitian form). Let the state vector for a localized quark-soliton bag state be denoted by $|B>$. Then

$$<B|\vec{P}|B>= 0, \quad \text{but} \quad <B|P^2|B> \ > \ 0.$$

Localized states contain spurious center-of-mass energy and center-of-mass fluctuational motion. The underlying translational invariance of the Lagrangian shows up as spurious states in the excitation spectrum.

In a non-relativistic theory, it is straightforward to construct the center-of-mass coordinate, which is just the mean of the quark positions if only (equal-mass) quarks are present, although it is complicated in practice to isolate the center-of-mass motion in a many-body system. In a relativistic field theory, the corresponding center-of-energy *operator* is not a tractable object. Pryce (1948) has discussed the merits and shortcomings of various candidates, and for finite mass interacting particles Krajcik and Foldy (1974) have given an explicit expansion in inverse powers of the constituent mass for an operator which satisfies the necessary relationships. One *approximate* center-of-energy operator is (Fokker, 1929; Pryce, 1948; Dethier, *et al.*, 1983)

$$\vec{R} \equiv \vec{K}/<H> = \int d^3r \, \vec{r} \, \mathcal{H}(\vec{r})/<H> , \qquad (4.2)$$

where the $\vec{K}$ are space-time infinitesimal generators of the Poincaré group and are associated with the boost operator, about which we will see more later.

There are 10 generators of the Poincaré group corresponding to infinitesimal translations and rotations in four-space. They are the operators of linear momentum $\vec{P}$, angular momentum $\vec{J}$, space-time rotation $\vec{K}$, and time translation H. They satisfy the commutation relations

$$[J_i, J_j] = i\epsilon_{ijk}J_k , \qquad (4.3\,a)$$

$$[J_i, P_j] = i\epsilon_{ijk}P_k , \qquad (4.3\,b)$$

$$[J_i, K_j] = i\epsilon_{ijk}K_k , \qquad (4.3\,c)$$

$$[K_i, P_j] = i\delta_{ij}H , \qquad (4.3\,d)$$

$$[K_i, K_j] = -i\epsilon_{ijk}J_k . \qquad (4.3\,e)$$

$$[K_i, H] = i\,P_i . \qquad (4.3\,f)$$

From (4.3) it follows that $\vec{R}$ satisfies the commutation relations

$$[R_i, P_j] = i\delta_{ij}\, H/<H> \qquad (4.4)$$

and

$$\frac{d\vec{R}}{dt} = i[H, \vec{R}] = \vec{P}/<H> , \qquad (4.5)$$

which are appropriate for a center-of-mass candidate; unfortunately, the components of $\vec{R}$ do not commute among themselves, but give rather

$$[R_i, R_j] = -i\epsilon_{ijk}J_k/<H>^2 , \qquad (4.6)$$

where J_k is the angular momentum operator. (4.4) is the condition that $\vec{R}$ be conjugate to $\vec{P}$, which is satisfied when acting on an eigenstate or when evaluated as in its expectation value. (4.5) is a "very pleasant" and non-trivial result. Recall that for the Dirac position operator $\vec{r}_D = \int d^3r\, \psi^\dagger \vec{r} \psi$ we have (for a local potential)

$$\frac{d\vec{r}_D}{dt} = \int d^3r\, \psi^\dagger \vec{\alpha} \psi\,, \tag{4.7}$$

which has eigenvalues equal in magnitude to the speed of light. Wave packets, however, satisfy (4.5) with $< H >= m$, and so does the position operator in the Foldy-Wouthuysen (1949) representation. The operator $\vec{R}$ is closely related to the F-W position operator.

4.2 Momentum Projection

The soliton coherent state described in the previous section is localized and so has no definite momentum. An eigenstate of momentum can be constructed from it by using the projection method of Peierls and Yoccoz (1957),

$$|\vec{p}>_1 = \int d^3X\, e^{i\vec{p}\cdot\vec{X}}|\vec{X}> = \int d^3X\, e^{i(\vec{p}-\vec{P})\cdot\vec{X}}|0>$$
$$= (2/pi)^3 \delta^3(\vec{p}-\vec{P})|0> \tag{4.8}$$

where we have used that $e^{-i\vec{P}\cdot\vec{X}}$ translates the state $|0>$ to $|\vec{X}>$. The subscript "1" denotes Peierls-Yoccoz. one of three projection methods to be discussed . The eigenvalue of $\vec{P}$ is denoted by $\vec{p}$, and $|\vec{X}>$ is a bag state localized about the point $\vec{X}$ and has the unnormalized form (say)

$$|\vec{X}> = \exp\left[\sum_{\vec{k}} \sqrt{\tfrac{1}{2}\omega_k}\, f_{\vec{k}}(\vec{X})\, a^{\pm}_{\vec{k}}\right] c_1^\dagger(\vec{X})\, c_2^\dagger(\vec{X})\, c_3^\dagger(\vec{X})|0>, \tag{4.9}$$

where $f_{\vec{k}}(\vec{X})$ is the Fourier coefficient of the σ field distribution [see Eqs. (3.45, 46)] centered at $\vec{X}$. It is simply related to the Fourier coefficient of the field centered at $\vec{X}=0$ by translation:

$$f_{\vec{k}}(\vec{X}) = e^{-i\vec{k}\cdot\vec{X}} f_{\vec{k}}(0)\,. \tag{4.10}$$

We have now committed ourselves to the plane wave representation for the σ and π operators (3.36). Note that $f_{-\vec{k}}(0) = f_{\vec{k}}^*(0)$. We will denote $f_{\vec{k}}(0)$ simply by $f_{\vec{k}}$. The operators $c_i^\dagger(\vec{X})$, i=1, 2, 3, create valence quark states, also centered at $\vec{X}$, with wave functions

$$\psi_i(\vec{r}, \vec{X}) = \psi_i(\vec{r} - \vec{X}). \tag{4.11}$$

There are well-known difficulties associated with Peierls-Yoccoz projection, as will be discussed in the next section.

The zero-momentum projected state is

$$|\vec{p}=0> = \int d^3X \, |\vec{X}> ; \tag{4.12}$$

no subscript on the state is used here because the various projection techniques discussed here agree on the zero momentum state.

The generalization to the single mode approximation for the sigma part of the state vector is straightforward. Since the sigma field is expanded in terms of plane waves, the operators are translationally invariant. The quark basis we have used is not translationally invariant but we neglect differences from unity of the overlaps of quark vacua centered on different points. This procedure cannot be exact but no obvious problems have arisen as yet. If it does cause trouble, one can always go to a plane-wave basis and work with a Dirac Hamiltonian projected onto positive-energy plane waves. A better procedure would be to calculate the distortion of the Dirac sea explicitly, see Chapter 9.

From (4.8) the expectation value of an operator $\mathcal{O}$ in the projected state is

$$<\mathcal{O}> = \frac{\int d^3X \, d^3X' <\vec{X}|\mathcal{O}|\vec{X}'>}{\int d^3X \, d^3X' <\vec{X}|\vec{X}'>}. \tag{4.13}$$

Let us introduce the mean and difference coordinates $\vec{Y} = \frac{1}{2}(\vec{X}+\vec{X}')$ and $\vec{Z} = \vec{X} - \vec{X}'$. Then, provided that $\mathcal{O}$ is translationally invariant,

$$<\mathcal{O}> = \frac{\int d^3Z <-\frac{1}{2}\vec{Z}|\mathcal{O}|\frac{1}{2}\vec{Z}>}{\int d^3Z <-\frac{1}{2}\vec{Z}|\frac{1}{2}\vec{Z}>}, \tag{4.14}$$

where the integration over $\vec{Y}$ cancels in the numerator and denominator. The integrand in the normalization is a product of σ and quark factors:

$$< -\tfrac{1}{2}\vec{Z} | \tfrac{1}{2}\vec{Z} > = N_\sigma(\vec{Z}) \, N_q(\vec{Z})^3 \,, \tag{4.15}$$

$$N_\sigma(\vec{Z}) = \exp\left[\tfrac{1}{2} \sum \omega_k f_{\vec{k}}^*(-\tfrac{1}{2}\vec{Z}) f_{\vec{k}}(\tfrac{1}{2}\vec{Z}) \right]$$
$$= \exp\left[\tfrac{1}{2} \sum \omega_k |f_{\vec{k}}|^2 \cos(\vec{k}\cdot\vec{Z}) \right] \,, \tag{4.16}$$

$$N_q(\vec{Z}) = \int d^3r \; \psi_0^\dagger(\vec{r} + \tfrac{1}{2}\vec{Z}) \; \psi_0(\vec{r} - \tfrac{1}{2}\vec{Z}) \,. \tag{4.17}$$

Here the subscript "0" refers to the $s_{1/2}$ quark states. The integrands for the expectation values of normal-ordered products of field operators are

$$< -\tfrac{1}{2}\vec{Z} | :\sigma(\vec{r})^n: | \tfrac{1}{2}\vec{Z} > = \overline{\sigma}(\vec{r}; \vec{Z})^n \, N_\sigma(\vec{Z}) \, N_q(\vec{Z})^3, \tag{4.18 a}$$

$$< -\tfrac{1}{2}\vec{Z} | :\pi(\vec{r})^n: | \tfrac{1}{2}\vec{Z} > = \overline{\pi}(\vec{r}; \vec{Z})^n \, N_\sigma(\vec{Z}) \, N_q(\vec{Z})^3, \tag{4.18 b}$$

where
$$\overline{\sigma}(\vec{r}; \vec{Z}) \equiv \tfrac{1}{2}\left[\sigma_0(\vec{r} - \tfrac{1}{2}\vec{Z}) + \sigma_0(\vec{r} + \tfrac{1}{2}\vec{Z}) \right] \,, \tag{4.19 a}$$

$$\overline{\pi}(\vec{r}; \vec{Z}) \equiv -i \sum \tfrac{1}{2}\omega_k \left[f_{\vec{k}}(\tfrac{1}{2}\vec{Z}) e^{i\vec{k}\cdot\vec{r}} - f_{\vec{k}}^*(-\tfrac{1}{2}\vec{Z}) e^{-i\vec{k}\cdot\vec{r}} \right] \,. \tag{4.19 b}$$

With these results we can now evaluate the nucleon mass in this approximation:

$$m = <H> = \frac{<\vec{P}=0|H|\vec{P}=0>}{<\vec{P}=0|\vec{P}=0>}, \tag{4.20}$$

where we remember that H is normal ordered.

4.3 Boost

To calculate other nucleon properties (*e.g.* electro-magnetic form factors) we need states with non-zero momentum. These could be constructed using finite-momentum projection. However, such a procedure has well-known difficulties: namely, the various states of good momentum are not related to each other by proper Galilean or Lorentz transformation (Peierls and Yoccoz, 1957; Griffin and Wheeler, 1957). That is to say, projection of a bag (or cluster) state, which is not the product of an intrinsic function times a center-of-mass function, leads to an intrinsic function which depends on the value of the momentum which is projected. Several solutions to this problem have been proposed. We follow here Lübeck, Birse, Henley and Wilets (1986) (see also Bardeen, *et al.*, 1975) where in order to circumvent the problem one operates on the zero-momentum state with the appropriate Lorentz boost operator to produce an approximate four-momentum eigenstate. The boosted state is defined by

$$|\vec{y}>_2 = e^{i\vec{y}\cdot\vec{K}}|\vec{p}=0>, \tag{4.21}$$

where the $\vec{K}$ operator, introduced by Eq. (4.2), is

$$\vec{K} = \int d^3r\; \vec{r}\, \mathcal{H}(\vec{r}). \tag{4.22}$$

with $\mathcal{H}(\vec{r})$ the Hamiltonian density. The quantity $\vec{y}$ is the rapidity in the direction of the velocity:

$$\vec{y} = \tfrac{1}{2}\hat{v}\ln\left(\frac{1+v}{1-v}\right) = \tfrac{1}{2}\hat{v}\ln\left(\frac{E+p}{E-p}\right), \tag{4.23}$$

with $E^2 = m^2 + p^2$. The subscript "2" in Eq. (4.21) refers to boost after projection onto a $p = 0$ state. Unless the state $|\vec{p} = 0>$ is an exact energy eigenstate, the state $|\vec{y}>$ is not an exact momentum eigenstate but it does lead to expectation values of energy and momentum with the correct Lorentz transformation properties. This can be seen as follows:

We have already seen that the operator $\vec{K}$ obeys the following commutation rules,

$$[K_i, P_j] = i\delta_{ij}H ,\qquad (4.3\,d)$$

$$[\vec{K}, H] = i\vec{P}, \qquad (4.3\,e)$$

We define

$$E(y) \equiv \,<\vec{y}|H|\vec{y}>, \qquad (4.24\,a)$$

$$p_\|(y) \equiv \,<\vec{y}|\hat{v}\cdot\vec{P}|\vec{y}> . \qquad (4.24\,b)$$

Then from (4.21) it follows that

$$\frac{dE}{dy} = p_\|(y) , \qquad (4.25\,a)$$

$$\frac{dp_\|}{dy} = E(y). \qquad (4.25\,b)$$

These differential equations have solutions (for the boundary conditions $E(0) = m$, $p_\|(0) = 0$)

$$E(y) = m\cosh(y) , \qquad (4.26\,a)$$

$$p_\|(y) = m\sinh(y), \qquad (4.26\,b)$$

where $m = E(0)$ is the expectation value of H in the $\vec{p} = 0$ state. This yields the Einstein relation

$$E^2 = m^2 + p^2. \qquad (4.27)$$

The boosted state is more conveniently labelled by the expectation value of the momentum, $\vec{p} = \hat{v}\, m\, \sinh(y)$. Then

$$|\vec{p}>_2 \equiv |\vec{y}(p)> . \qquad (4.28)$$

The use of boost to calculate magnetic moments and form factors will be described elsewhere.

4.4 Variation After Projection

It is well known in nuclear physics that energy variation before projection can be a dangerous procedure. In the present case, the expectation value of the energy depends upon the functions

$$u(r), \quad v(r), \quad \sigma_0(r) \text{ [or } f_k], \quad F_m \text{ and } \omega_k,$$

where u and v are the upper and lower components of $\psi_0(\vec{r})$. The coefficients F_m appear only in the case of the general single mode. A full variation with respect to all of these functions appears to be prohibitive at present. Instead, the *forms* obtained from solving the mean field equations, now denoted by a tilde, were utilized, and scaling was introduced as follows:

$$\sigma_0(r) - \sigma_v = \xi[\tilde{\sigma}_0(r/\lambda) - \sigma_v] \tag{4.29 a}$$

$$u(r) = \tilde{u}(r/\delta), \qquad v(r) = \gamma\tilde{v}(r/\delta), \tag{4.29 b, c}$$

where ξ, λ, δ, and γ are dimensionless variational parameters. The normalization of ψ_0 must, of course, be readjusted. An alternative parametrization of σ_0, using a Fermi function form,

$$\sigma_0(r) = \left\{ \sigma_v - \frac{\xi}{1 + \exp\left[(r - r_0)/\mu\right]} \right\}, \tag{4.30}$$

was also tried. This allows independent variations of the surface thickness μ and radius r_0 in $\sigma_0(r)$.

As already noted. the commutation relations (3.37) are satisfied for any set of ω_k. Another choice, say Ω_k, defines operators $A_{\vec{k}}$ which are linear combinations of $a_{\vec{k}}$ and $a_{\vec{k}}^\dagger$, and can be regarded as a Bogoliubov transformation on the creation and annihilation operators:

$$A_{\vec{k}} = \tfrac{1}{2}\left(\sqrt{\frac{\Omega_k}{\omega_k}} + \sqrt{\frac{\omega_k}{\Omega_k}}\right) a_{\vec{k}} + \tfrac{1}{2}\left(\sqrt{\frac{\Omega_k}{\omega_k}} - \sqrt{\frac{\omega_k}{\Omega_k}}\right) a_{-\vec{k}}^\dagger. \tag{4.31}$$

Such a transformation is well-defined as long as no Ω_k vanishes.

The corresponding "vacuum" state $|\Omega>$, defined by

$$A_{\vec{k}}|\Omega> = 0, \qquad \text{for all} \quad \vec{k}, \qquad (4.32)$$

is a Gaussian wave packet in functional space. Its principal axes are given by the plane-wave expansion functions, and its widths are $\Omega_{\vec{k}}^{-\frac{1}{2}}$. The special choice

$$\omega_k^2 = m_\sigma^2 + k^2 \qquad (4.33)$$

with

$$m_\sigma^2 = m_{GB}^2 = U''(\sigma_v) \qquad (4.34)$$

diagonalizes the Hamiltonian for small oscillations about the mean σ field in the absence of quarks and minimizes the energy of the corresponding vacuum state.

The coherent state used to describe the nucleon is a displaced Gaussian wave packet of the same form as $|\Omega>$, but centered on the classical field configuration $\sigma_0(r)$. The general single-mode state allows for replacement of this Gaussian by a more general function for the chosen single mode.

As already noted, a better starting point for the nucleon state would be to expand the σ field in distorted waves (see Chapter 9). However, we are here committed to a plane-wave basis in order to facilitate projection. In choosing a coherent state with $\Omega_k \neq \omega_k$ to describe the nucleon. we also change the vacuum for the nucleon, and this produces an infinite shift in the vacuum contribution to the energy. The procedure is to calculate only the (finite) differences in energies, and other properties, between the nucleon and the (un-observable) vacuum state, $|\Omega>$. In order to focus on the quantum fluctuations in the vicinity of the nucleon, the variational princi-ple is invoked for the difference in energy between the nucleon and the vacuum. Since the energy of each eigenstate is stationary with respect to arbitrary variations. so also is the energy difference, at least for independent variations of the parameters used to describe both states; Ω_k is constrained to be the same in both states. To test whether this procedure is meaningful, certain virial theorems, described next, were utilized. As noted below, optimization with respect to the Ω_k leads to significant improvement in the satisfaction of the virial theorems.

4.5 Virial Theorems

The time derivatives of the expectation value of any time-independent operator vanish in a stationary eigenstate of the Hamiltonian:

$$\frac{d}{dt} <O> = i < [H, O] > = 0 . \tag{4.35}$$

Consider the following operators:

$$O_1 \equiv \int \psi^\dagger(\vec{r}) \, \vec{r} \cdot \vec{p} \, \psi(\vec{r}) \, d^3r \tag{4.36 a}$$

$$O_2 \equiv \int \pi(\vec{r}) \, d^3r \tag{4.36 b}$$

$$O_3 \equiv \int :\sigma(\vec{r}) \, \pi(\vec{r}): \, d^3r . \tag{4.36 c}$$

The commutators of these operators with respect to the (normal ordered) Hamiltonian lead to corresponding "virial theorems", which require the vanishing of

$$V_1 \equiv \int d^3r < \psi^\dagger(\vec{r}) [\, \vec{\alpha} \cdot \vec{p} - g\,\beta\,\vec{r} \cdot \vec{\nabla} \sigma(\vec{r})\,] \psi(\vec{r}) > / < H > \tag{4.37 a}$$

$$V_2 \equiv \int d^3r < -\nabla^2 \sigma(\vec{r}) + U'(\sigma(\vec{r})) + g\overline{\psi}(\vec{r})\psi(\vec{r}) > \tag{4.37 b}$$

$$V_3 \equiv \int d^3r <: \left[-\pi(\vec{r})^2 + \sigma(\vec{r}) \frac{\partial H}{\partial \sigma(\vec{r})} \right] :> / < H > . \tag{4.37 c}$$

(The factors $< H >^{-1}$ are introduced in (a) and (c) to render all of the V_i dimensionless.) In the first two cases, commutation with

H does not destroy normal ordering while in the last case it does. Eq. (4.37c) contains a normal-ordered piece plus divergent terms. However, with $\omega_k = \sqrt{m_{GB}^2 + k^2}$, the divergent terms cancel leaving only the finite part. Nevertheless, the divergent part is ignored even when $\omega_k \neq \sqrt{m_{GB}^2 + k^2}$.

The first is a generalization of the familiar virial theorem of non-relativistic quantum mechanics. In that case, it is equivalent to energy minimization with respect to the scale of length for the wave function. The second is the integral of the field equation for $-i\dot{\pi}$. The third is the σ-field analogue of the first.

Each of these quantities vanishes in the mean field approximation if one has obtained self-consistent solutions. None of these quantities vanishes automatically in the projected state. The satisfaction of these quantities after variation is a necessary, but obviously not sufficient, condition for a good solution to the field equations.

4.6 Numerical results

Projection of a bag state removes spurious center-of-mass energy. In Fig. 4.1 is a comparison of the energy of a projected bag state with the energy obtained by subtraction of the mean-square momentum, *i.e.* $(<H>^2 - <P^2>)^{1/2}$, for one family of parameters (b^2/ac) and varying c. Note that the two curves cross at $c \simeq 3,000$.

A good test for the quality of a wave function is the satisfaction of the virial theorems. They have been rendered dimensionless, and should be small compared with unity. Energy minimization alone only gives a number with no reference for comparison. For the virial theorems, it was essential to include variation of the ω_k, which was effected by varying the m_σ in Eq. (4.33) and including an exponential cut-off parameter, that is,

$$m_\sigma^2 = m_{GB}^2 + \delta m^2 \, e^{-\alpha k}. \tag{4.38}$$

It does not destroy normal ordering while in the last case it does ... [text partially obscured] ... contains a normal-ordered piece plus divergent terms. However, within ... [text partially obscured] ... the divergent terms cancel leaving ... [text partially obscured]

... [text partially obscured] ...

... [text partially obscured] ...

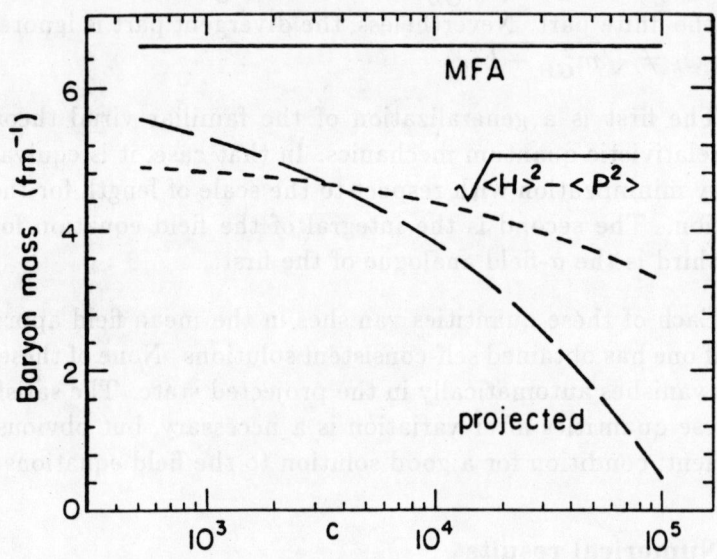

Figure 4.1. Energy of a projected nucleon bag as a function of the model parameter c, compared with subtraction of the spurious center-of-mass momentum. (Lübeck, *et al.*, 1986)

As can be seen in Fig. 4.2, the three virials cross zero *near* the same value of δm^2 and near the (very flat) minimum in $<H>$. This provides a justification for the procedure of varying M_σ, even though the vacuum energy is "disturbed" in the process.

The actual change in energy due to projection and variation is large but not very interesting in itself, since the parameters are readjusted to fit key data, such as masses and sizes. What is interesting is the quantitive fit to other data compared with the mean field approximation. Lübeck *et al.* (1986) found significant improvement in reproducing the values of the nucleon magnetic moments, the axial coupling g_A, and the pionic mass.

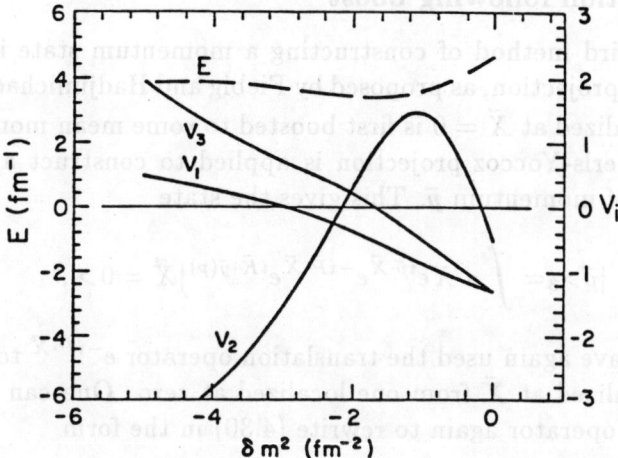

Figure 4.2. Dependence of the nucleon projected energy and virial theorems on the parameter δm^2. (Lübeck, *et al.*, 1986)

There are (at least) two methods for calculating the nucleon magnetic moment. One is the expectation value of the magnetic moment operator in the $\vec{P} = 0$ state,

$$\mu(\text{static}) = \left\langle \int d^3 r (\vec{r} \times \vec{j}(\vec{r}))_z \right\rangle_{m=\frac{1}{2}}. \qquad (4.39\,a)$$

The other is obtained by boosting the rest state to momentum $\pm\vec{q}/2$, and calculating the matrix element of the current linear in $\vec{q}$:

$$\mu(\text{boost}) = \frac{1}{2Mq}\left\langle -\tfrac{1}{2}q \right| \left(\vec{K} \times \vec{j}(0) \right)_{z,+} \left| \tfrac{1}{2}q \right\rangle_{m=\frac{1}{2}}, \qquad (4.39\,b)$$

where the subscript "+" indicates an averaging over both orderings of $\vec{K}$ and $\vec{j}(0)$. The values so obtained would agree if the state vectors were energy eigenstates. The degree of agreement is another measure of the quality of the wave functions. Lübeck *et al.* find agreement to a few percent. Further numerical results are presented in Chapter 7.

4.7 Projection following boost

The third method of constructing a momentum state is boost followed by projection, as proposed by Fiebig and Hadjimichael (1984). A state localized at $\vec{X} = 0$ is first boosted to some mean momentum $\vec{p}$; then Peierls-Yoccoz projection is applied to construct an exact eigenstate of momentum $\vec{p}$. This gives the state

$$|\vec{p}>_3 = \int d^3X e^{i\vec{p}\cdot\vec{X}} e^{-i\vec{P}\cdot\vec{X}} e^{i\vec{K}\cdot\vec{y}(p)} |\vec{X} = 0>, \qquad (4.40)$$

where we have again used the translation operator $e^{-i\vec{P}\cdot\vec{X}}$ to obtain a state localized at $\vec{X}$ from one localized at zero. One can use the translation operator again to rewrite (4.30) in the form

$$|\vec{p}>_3 = \int d^3X e^{i\vec{p}\cdot\vec{X}} \left[e^{-i\vec{P}\cdot\vec{X}} e^{i\vec{K}\cdot\vec{y}(p)} e^{i\vec{P}\cdot\vec{X}} \right] |\vec{X}>. \qquad (4.41)$$

Although this is a state of good momentum, it is not guaranteed to satisfy Lorentz invariance nor, in particular, the Einstein relation, $E^2 = m^2 + p^2$ because various components of the $\vec{X} = 0$ state can contribute in the boosted state with the same momentum $\vec{p}$.

To summarize the properties various momentum states, $|\vec{p}>_1$ and $|\vec{p}>_3$ have precise momenta, but do not satisfy the Einstein relation. $|\vec{p}>_2$ is not an exact momentum eigenstate, but does satisfy the Einstein relation. Since both $|\vec{p}>_2$ and $|\vec{p}>_3$ are obtained by boost (whereas $|\vec{p}>_1$ is not), they should be "good" functions. We should note, however, that if $|\vec{p} = 0>$ is an exact eigenstate of the energy, then $|\vec{p}>_2$ is an exact four-momentum eigenstate; this is not necessarily true of $|\vec{p}>_3$.

Chapter five

THE GENERATOR COORDINATE METHOD

5.1 The Hill-Wheeler integral equation

Large amplitude collective motion requires techniques beyond normal mode or RPA constructions. The method of generator coordinates (GCM) (Hill and Wheeler, 1953; Griffin and Wheeler, 1958) was first introduced to describe collective dynamics of nuclear systems, such as rotation and vibration, in order to avoid the troublesome problem of redundant coordinates. It may be regarded as an expansion in a complete set of states. Because of the further degrees of freedom contained in weight functions, the expansion is overcomplete. In practice, the set is always truncated so that overcompleteness is not a problem in principle; however, lack of orthogonality can be a delicate numerical problem. The efficiency of the expansion is governed by the choice of a good, physical, set of states so that the size of the set is small. A variational principle leads to the Hill-Wheeler integral equation for the weight functions. In special cases, symmetry dictates the choice of the weight functions: For translation, the functions are plane waves and for rotations they are Wigner D functions; in these cases, the GCM is equivalent to Peierls-Yoccoz (1957) projection.

The GCM has also been considered for large amplitude bag dynamics (Wilets, 1985), with particular application to N-N scattering (Schuh, 1985; Crawford and Miller, 1987), π-N coupling (Dethier, 1985) and $N\overline{N}$ reactions (Crawford et al., 1985). The method is especially useful there because of the difficulty in otherwise defining collective variables in a relativistic field theory.

We begin with a very general description of the method. Consider a parameter or set of parameters $\{\alpha\}$ which describe the static configuration of a system of quarks and the soliton field and let $|\alpha, n>$ denote a set of basis states which is complete for any α. These may be obtained, for example, by constrained mean field calculations, as will be described later. The GCM state vector is written

$$|\Psi; gc> = \sum_n \int \phi_n(\alpha) |\alpha, n> d\alpha. \qquad (5.1)$$

Since the set is complete for each α, the expansion, as noted above, is overcomplete. In practice, this causes no problem since the sum is truncated to a small number of terms. In what follows, we consider only a single term and suppress n. (Alternatively, ϕ may be regarded as a vector with components ϕ_n. Similarly, α can be a set of parameters, including angles.) In practical cases, several configurations may be required. The generalization is straightforward.

The weight function $\phi(\alpha)$ is obtained by extremizing the expectation value of the Hamiltonian

$$\int d\alpha \phi^*(\alpha) < \alpha |H| \alpha' > \phi(\alpha') d\alpha'$$

subject to the normalization constraint

$$\int d\alpha' < \alpha |H - E| \alpha' > \phi(\alpha') = 0. \qquad (5.2)$$

This is the basic Hill-Wheeler GCM integral equation for $\phi(\alpha)$. Depending upon whether the spectrum is discrete or continuous, it is either an eigenvalue or a scattering equation. Because it is a homogeneous integral equation, it is notoriously unstable numerically.

5.2 The H-W differential equation

Although one can work with the integral equation (which does, however, require regularization, *c.f.* Wampler and Wilets, 1988), it is instructive and useful to consider the *approximate* differential equation which it satisfies. The differential equation is free of numerical instabilities. For a system which has well developed collective motion, we expect $< \alpha | H - E | \alpha' >$ to fall off rapidly as a function of $\alpha - \alpha'$. To utilize that property, it is convenient to introduce the mean and relative parameters

$$\bar{\alpha} = \tfrac{1}{2}(\alpha + \alpha'), \qquad \delta = \alpha - \alpha'. \tag{5.3}$$

Then

$$< \Psi; gc | H - E | \Psi; gc >$$

$$= \int d\bar{\alpha} \int d\delta \, \phi^*(\bar{\alpha} + \tfrac{1}{2}\delta) < \bar{\alpha} + \tfrac{1}{2}\delta | H - E | \bar{\alpha} - \tfrac{1}{2}\delta > \phi(\bar{\alpha} - \tfrac{1}{2}\delta)$$

$$= \int d\bar{\alpha} \int d\delta \, [\phi^*(\bar{\alpha}) + \tfrac{1}{2}\delta \, \phi'^*(\bar{\alpha}) + \tfrac{1}{8}\delta^2 \phi''^*(\bar{\alpha}) + \cdots]$$

$$\times < \bar{\alpha} + \tfrac{1}{2}\delta | H - E | \bar{\alpha} - \tfrac{1}{2}\delta >$$

$$\times [\phi(\bar{\alpha}) - \tfrac{1}{2}\delta \phi'(\bar{\alpha}) + \tfrac{1}{8}\delta^2 \phi''(\bar{\alpha}) + \cdots]. \tag{5.4}$$

If $< \bar{\alpha} + \tfrac{1}{2}\delta | H - E | \bar{\alpha} - \tfrac{1}{2}\delta >$ is even in δ, the only integrals which survive are of the form

$$O_n(\bar{\alpha}) \equiv \int < \bar{\alpha} + \tfrac{1}{2}\delta | O | \bar{\alpha} - \tfrac{1}{2}\delta > \delta^n d\delta \tag{5.5}$$

for n even. Here the operator O is either H or $N \equiv 1$. Through order ϕ'' and $|\phi'|^2$, we have

$$< \Psi; gc | H - E | \Psi; gc >$$

$$= \int d\bar{\alpha} \left[\phi^*(H_0 - E N_0)\phi + \tfrac{1}{4}(H_2 - E N_2)(-\phi'^* \phi' + \tfrac{1}{2}\phi''^* \phi + \tfrac{1}{2}\phi^* \phi'') \right]. \tag{5.6}$$

This may be cast into the more familiar form

$$\int \tilde{\phi}^* \left(-\frac{d}{d\bar{\alpha}} \frac{1}{2B(\bar{\alpha})} \frac{d}{d\bar{\alpha}} + V(\bar{\alpha}) - E \right) \tilde{\phi} \, d\bar{\alpha} \tag{5.7}$$

with

$$V(\overline{\alpha}) = \frac{H_0}{N_0} + N_0^{-\frac{1}{2}} \frac{d}{d\alpha}(H_2 - E\,N_2)\frac{d}{d\alpha}N_0^{-\frac{1}{2}} + \frac{1}{8\,N_0}\frac{d^2}{d\overline{\alpha}^2}(H_2 - E\,N_2),$$

$$(5.8\,a)$$

$$B(\overline{\alpha}) = -\frac{N_0}{H_2 - E\,N_2} \qquad\qquad (5.8\,b)$$

and

$$\tilde{\phi} = N_0^{\frac{1}{2}}\phi. \qquad\qquad (5.8\,c)$$

Note the explicit dependence of B on E. This can be removed by a more general definition of $\tilde{\phi}$ in terms of ϕ which contains ∇^2, see Ring and Schuck (1980), Achtzehnter (1988).

We can define α such that $\alpha \to r$ as $r \to \infty$, where r is (say) the separation of two bags. However, $V(\overline{\alpha})$ has no simple intuitive interpretation as a potential until $B(\overline{\alpha})$ is determined!

Following Fujiwara (1959), we introduce a change of variable from $\overline{\alpha}$ to x (which is the same as r for a head-on collision) such that

$$x = X(\overline{\alpha}) = \int^{\overline{\alpha}}[B(\alpha')/\mu]^{\frac{1}{2}}d\alpha' = -\int_{\overline{\alpha}}^{\infty}\left[\left(\frac{B(\alpha')}{\mu}\right)^{\frac{1}{2}} - 1\right]d\alpha' + \overline{\alpha},$$

$$(5.9\,a)$$

$$\tilde{\phi}(\overline{\alpha}) = \left(\frac{B(\overline{\alpha})}{\mu}\right)^{\frac{1}{4}}\psi(x). \qquad\qquad (5.9\,b)$$

where $\mu = m/2$ is the reduced nucleon mass. Then variation of Eq.(5.7) with respect to ψ^* yields

$$\left(-\frac{1}{2\mu}\frac{d^2}{d\,x^2} + V + V_1 - E\right)\psi(x) = 0 \qquad (5.10)$$

with

$$V_1 = -\frac{B''}{8\,B^2} + \frac{7(B')^2}{32\,B^3}. \qquad\qquad (5.11)$$

This is consistent with $B \to \mu$ as $\alpha \to x$ (or r). The generalization to three dimensions is straightforward. Note that V_1 is also energy-dependent in this formulation.

5.3 Example: the N-N interaction

The pioneering calculations on the N-N interaction in the context of the MIT bag were performed by DeTar (1978). These were static calculations of quarks interacting through one gluon exchange in a deformed cavity. While instructive, one cannot interpret the results in terms of an N-N potential until one has well-defined dynamics.

The equation for the dynamical collision of two bags has been calculated by Schuh *et al.* (1985), but without gluon exchange and for zero impact parameter. The state vector $|\alpha>$ is determined by extremizing the expectation value of the total Hamiltonian, $<\alpha|H|\alpha>$, with respect to a variational mean field wave function for the quarks and a coherent state wave function for the soliton field subject to a constraint

$$<\alpha|Q|\alpha> = Q_0 \qquad (5.12)$$

where Q is some moment of the quark distribution

$$Q = \int \overline{\psi}\, q(\vec{r})\, \psi \, d^3r. \qquad (5.13)$$

The constrained mean field equations now assume the form

$$\{\vec{\alpha}\vec{p} + \beta[g\sigma_0(\vec{r}) - \lambda\, q(\vec{r})] - \epsilon_n\}\psi_n = 0, \qquad (5.14\,a)$$

$$-\nabla^2\sigma_0 + U'(\sigma_0) + g \sum_{n(valence)} \overline{\psi}_n\psi_n = 0. \qquad (5.14\,b)$$

where λ is a Lagrange multiplier. Instead of specifying the constraint function $q(\vec{r})$ explicitly and solving the pair of equations (5.14 a, b) simultaneously and selfconsistently, it is actually more physical and simpler to specify the function in square brackets

$$g\sigma_0(\vec{r}) - \lambda\, q(\vec{r}) \equiv \mathcal{V}(\vec{r}). \qquad (5.15)$$

This plays the role of a scalar generating potential for the quarks. $\mathcal{V}(\vec{r})$ is determined by folding the volume formed by the union (for

$\alpha > 0$) or the intersection (for $\alpha < 0$) of two spheres, whose centers are separated by a distance $|\alpha|$, with a Yukawa form factor. $\alpha > 0$ corresponds to prolate deformations and $\alpha < 0$ to oblate deformations. See Fig. 5.1.

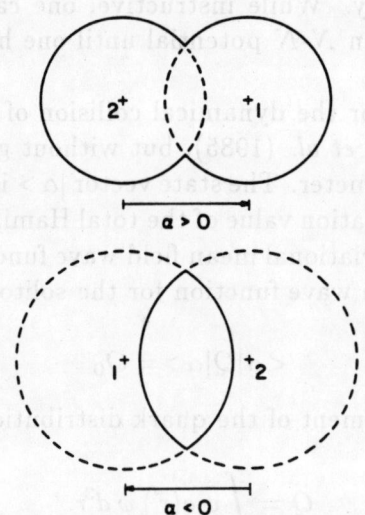

Figure 5.1. The geometry for determining $\mathcal{V}$. The volume enclosed by the solid line is folded with a Yukawa form factor. Fig. a is for $\alpha > 0$, Fig. b is for $\alpha < 0$.

The static quark densities for various deformations are shown in Figs. 5.2.

In the calculations of Schuh et al., the reduction to a Schrödinger equation differs slightly from that described above. The mass parameter $B(\overline{\alpha})$ is strongly $\overline{\alpha}$-dependent, as is shown in Fig. 5.3. The resulting $x(\overline{\alpha})$ is shown in Fig. 5.4, and the effective potential $V_{eff} = V + V_1$ is shown in Fig. 5.5. For $x \to -\infty$, $V_{eff} \to 6g\sigma_v \approx 4$ GeV. For the zero-impact parameter trajectory, the boundary condition on $\psi(x)$ is $\psi(-\infty) = 0$. V_{eff} is similar to the "static" energy $<\alpha|H|\alpha>$ at large x, but at small x turns up displaying the onset of a repulsive core. This occurs because the mass parameter becomes very small leading to high kinetic energy. The results clearly indicate the importance of dynamics in the scattering process: static energies do not tell the story.

Of special interest to many researchers who work with quark structures may be that this provides a method to construct quark wave functions for all stages of hadron-hadron collisions.

Crawford (1986) has approached the nucleon-nucleon scattering problem by dividing space into an inner soliton bag region and an outer N-N potential region, with the appropriate continuity conditions at the boundary. The inner region is described by six quarks in a deformed sigma field. The quark wave functions are approximated by "sudden" or "frozen" displaced three quark clusters, in contrast to the adiabatic quark states of Schuh *et al.* The quarks interact through an effective contact potential to simulate one gluon exchange. The strength of the interaction and the matching radius are adjustable parameters. The outer region is described through the use of a Reid potential. Quite satisfactory phase shifts are obtained.

5.4 Molecular and cluster model wave functions

The adiabatic states employed by Schuh are an example of what are frequently called molecular states in analogy with homopolar diatomic molecular functions. They contain both reflectional and axial symmetry. The problem of classifying and constructing such states in both the current (*i.e.* soliton) and constituent (*i.e.* potential, cluster) models has been discussed in detail by Stancu and Wilets (1987, 1988). A schematic representation of the single particle eigenvalues is depicted in Fig. 5.6 for configurations which vary from the united bag or cluster configuration ($Z = 0$) to the separated ($Z = \infty$) configuration. Note that the separated bags are doubly degenerate with respect to right and left states, or with repect to the parity referred to the center of the combined system.

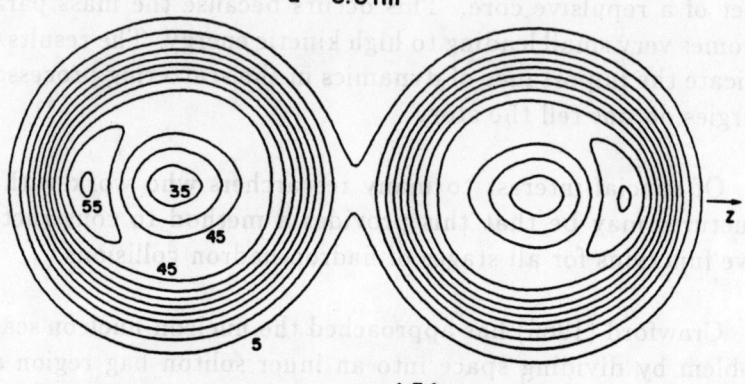

α = 3.0 fm

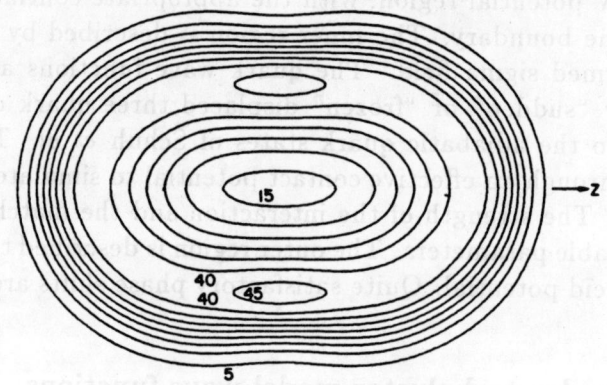

α = 1.5 fm

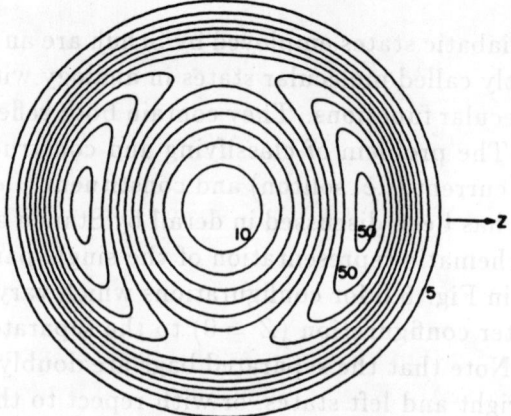

α = 0

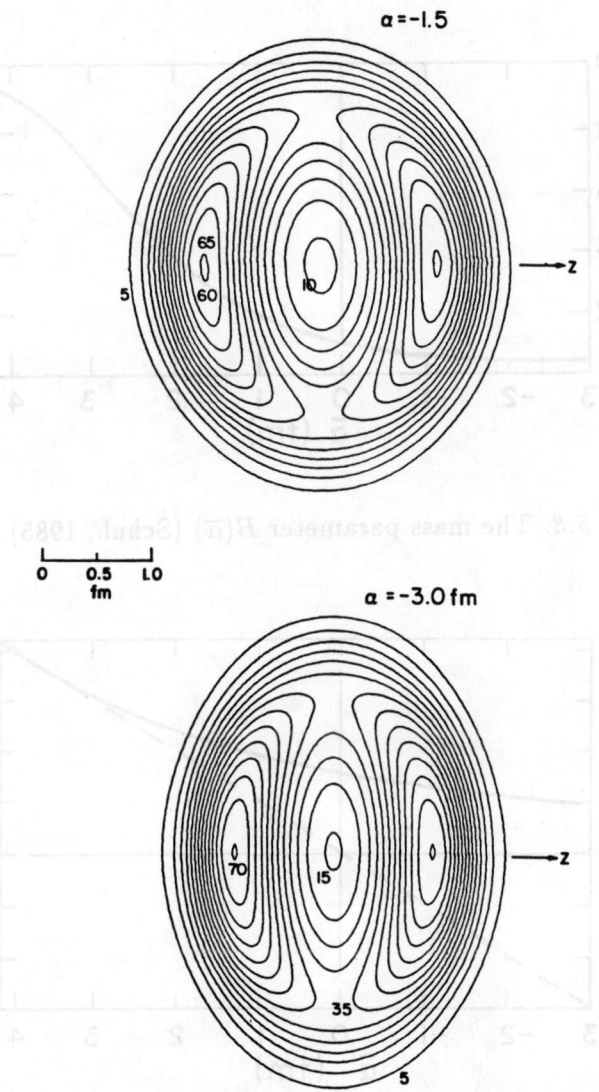

Figures 5.2. Quark densities obtained for various deformations (Schuh, 1984). The symmetry axis, z, is horizontal. Contours are labelled every 5 (arbitrary) units. The orbital filling used contains both even and odd states.

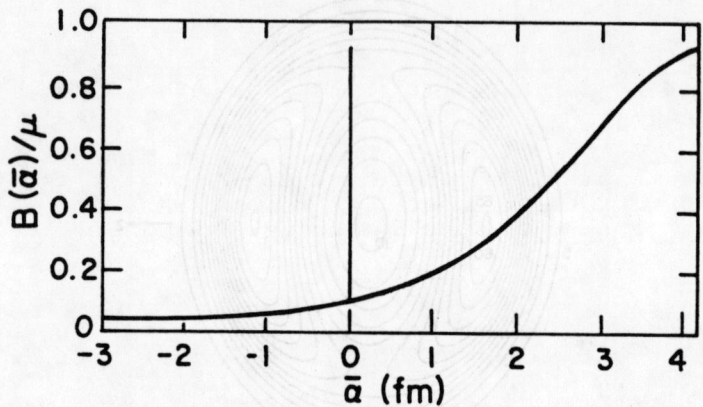

Figure 5.3. The mass parameter $B(\overline{\alpha})$ (Schuh, 1985).

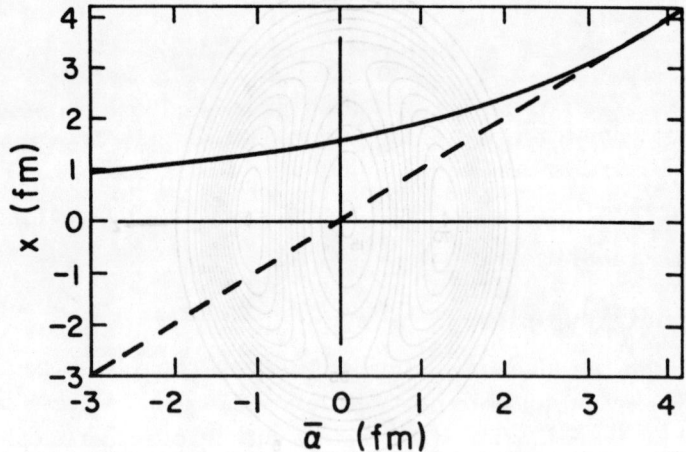

Figure 5.4. Rescaled length parameter $x(\overline{\alpha})$ (Schuh, 1985). Note that at the "spherical" configuration $\alpha = 0$, the rescaled separation parameter is finite: $x \approx 1.6$ fm.

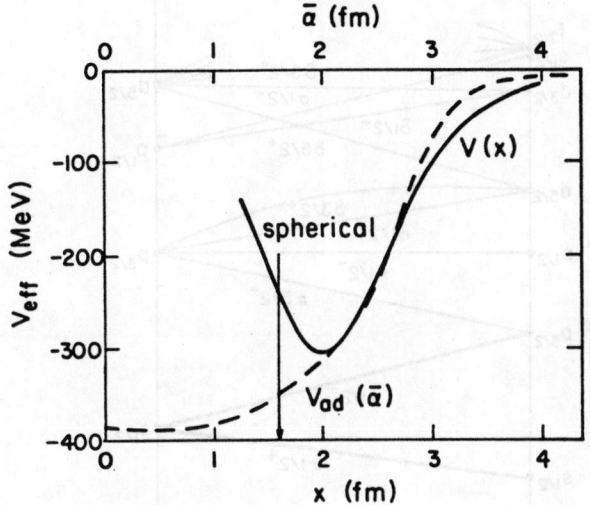

Figure 5.5. The "effective potential" $V + V_1$ as a function of x (solid curve), and $V_{ad} =< \overline{\alpha}|H|\overline{\alpha}>$ as a function of $\overline{\alpha}$ (dashed curve), at zero energy. (Schuh, 1985)

Calculations in the cluster model have usually been executed using the product of three right and three left orbitals. In the paper just referred to, it is shown that this neglects important configurations which have important effects on the energy, at least for small inter-cluster separations.

5.5 Internal dynamics

Although the GCM as described above is capable, in principle, of providing a complete dynamical description, the process can be expedited by providing the state vectors with a collective boost (Haff and Wilets, 1973). The following is a generalization of the boost procedure of Sec. 4.3 to the case of nonuniform motion.

Let us suppose (as in Sec. 5.3) that $|\alpha >$ is obtained by extremizing the expectation value of the Hamiltonian subject to the constraint that for some operator Q has a prescribed value

$$<Q>= Q_0 . \qquad (5.16)$$

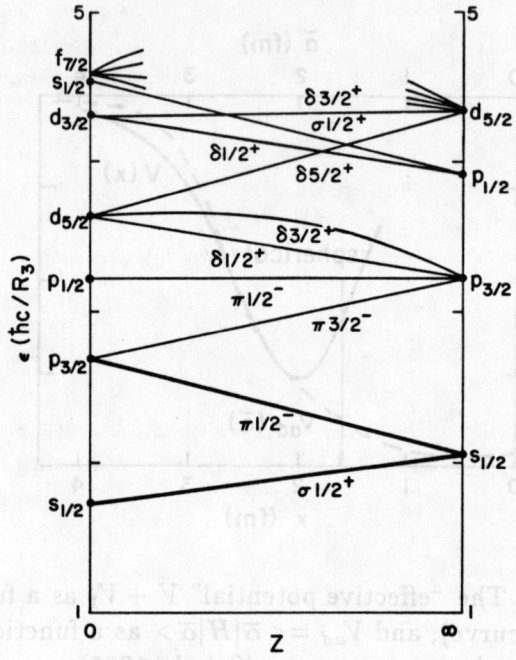

Figure 5.6. Schematic representation of the eigenenergies of molecular-type orbital as a functions of a separation parameter Z (Stancu and Wilets, 1987).

This is equivalent to extremizing the expectation value of

$$\hat{H} = H - \lambda Q, \tag{5.17}$$

where λ is a Lagrange multiplier.

We now consider the added constraint that the time derivative of Q has a prescribed value:

$$<\dot{Q}> = \dot{Q}_0. \tag{5.18}$$

This is equivalent to extremizing the expectation value of

$$\hat{\hat{H}} = H - \lambda Q - \beta \dot{Q}. \tag{5.19}$$

Now $\dot{Q} = i[H, Q] = i[\hat{H}, Q]$. Let

$$\hat{H}|\alpha> = \hat{E}(\alpha)|\alpha>. \tag{5.20}$$

In general, α can be used generally to denote λ or Q_0 or any parameter that describes the state. We use it here to denote Q_0. Then

$$
\begin{aligned}
|\alpha, \beta> &\approx \{1 - (\hat{E}(\alpha) - \hat{H})^{-1}\beta\, i\, [\hat{H}, Q]\}|\alpha> \\
&\approx \{1 - (\hat{E}(\alpha) - \hat{H})^{-1}\beta\, i\, (\hat{H}Q - Q\hat{E}(\alpha))\}|\alpha> \\
&\approx e^{i\beta Q}|\alpha> .
\end{aligned} \tag{5.21}
$$

We now show that β can be given a semi-classical interpretation as the momentum conjugate to Q_0. In the following, we use the notation $<X> \equiv <\alpha, \beta|X|\alpha\beta>$. By Feynman's theorem,

$$\frac{\partial <\hat{H}>}{\partial \beta} = - <\dot{Q}>, \tag{5.22}$$

but also

$$\frac{\partial <\hat{H}>}{\partial \beta} = \frac{\partial <\hat{H}>}{\partial \beta} - <\dot{Q}> - \frac{\partial <\dot{Q}>}{\partial \beta}\beta, \tag{5.23}$$

or

$$\frac{\partial <\hat{H}>}{\partial \beta} = \frac{\partial <\dot{Q}>}{\partial \beta}\beta, \tag{5.24}$$

Let us define

$$\left.\frac{\partial <\dot{Q}>}{\partial \beta}\right|_{\beta=0} \equiv \frac{1}{B} . \tag{5.25}$$

The energy must be even in the "velocity" $<\dot{Q}>$ or in the "momentum" β. Therefore, from (5.25) we have to order β^3

$$<\dot{Q}> = \frac{\beta}{B} = \frac{\partial}{\partial \beta}\left(\frac{\beta^2}{2B}\right) \tag{5.26}$$

which may be regarded as the Hamilton equation $\dot{Q} = \partial H/\partial P$. Treating $< Q >= Q_0 \equiv \alpha$ and $< \dot{Q} >= \beta/B = \dot{Q}_0$ as classical conjugate variables, we quantize by setting

$$\beta = -i\,\partial/\partial Q_0 \equiv i\,\partial/\partial\alpha. \qquad (5.27)$$

Thus we can now write a modified GCM wave function:

$$|\Psi; gc>= \int e^{-Q\partial/\partial\alpha}\phi(\alpha)|\alpha>, \qquad (5.28)$$

where it is to be understood that the differential $\partial/\partial\alpha$ acts only on the weight function $\phi(\alpha)$ and not on the state vector $|\alpha>$.

A special case is translation. Comparison of (5.21) with (4.21) and the identification $B = m$ and $\vec{y} \approx \vec{v} \approx \vec{p}/m$ suggests the further identification of $\vec{K}/m$ as the center-of-mass operator. This is a useful approximation, but is certainly not exact (Pryce, 1948; Krajcik and Foldy, 1974; Dethier, Goldflam, Henley and Wilets, 1983).

The Fiebig-Hadjimichael state $|\vec{p}>_3$ affords a prototypical way to introduce generator coordinates in a relativistic field theory for the case of translational motion. We can construct a wave packet by multiplying $|\vec{p}>_3$ of Eq. (4.41) by a momentum weight factor $\tilde{\phi}(\vec{p})$ and integrating over $\bar{d}^3 \equiv d^3p/(2\pi)^3$. Through order p^2, we can further replace the rapidity $\vec{y}$ by $\vec{p}/m$, and replace $\vec{p}$ by $-i\vec{\nabla}$ operating on $e^{i\vec{p}\cdot\vec{X}}$. Then with the definition

$$\phi(\vec{X}) = \int \bar{d}^3p\,\tilde{\phi}(\vec{p})e^{i\vec{p}\cdot\vec{X}} \qquad (5.29)$$

for the weight function, the GCM wavefunction can be written

$$|\Psi; gc>= \int d^3X \left[e^{-i\vec{P}\cdot\vec{X}}e^{\vec{K}\cdot\vec{\nabla}/m}e^{i\vec{P}\cdot\vec{X}} \right] \phi(\vec{X})|\vec{X}> \qquad (5.30)$$

where it is understood that $\vec{\nabla}$ operates only on $\phi(\vec{X})$ and not on $|\vec{X}>$. We now apply this wavefunction to the problem of Dirac phenomenology.

5.6 Example: the nucleon as a composite

There has been much success in Dirac phenomenology, where the nucleon is described by a Dirac equation with external isoscalar, Lorentz scalar and vector potentials, both in scattering (Clark *et al.*, 1983; McNeil *et al.*, 1983; Shepard *et al.*, 1983, Machleidt, Holinde and Elster, 1987) and many body problems (Brown *et al.*, 1970; Wilets,1979; Walecka, 1974; Anastasio *et al.*, 1983; Serot and Walecka, 1986, ter Haar and Malfliet, 1987). This has been deemed suspect by some (c.f. Brodsky, 1984) because the Dirac equation describes point particles devoid of internal structure and of capability for intrinsic excitation, whereas the nucleon has a size (e.g. the proton charge form factor) of more than four times its Compton wave length and a rich spectrum. In order to test the validity of the Dirac approach for external fields, Achtzehnter and Wilets (Achtzehnter, 1988; Achtzehenter and Wilets, 1988) constructed a Pauli (two-component Schrödinger) equation and compared it term-by-term with the equivalent Foldy-Wouthuysen (1950) transformed Dirac equation.

Although the problem was formulated in the framework of the soliton model, the results are model-insensitive within the basic assumption that the external fields interact only with valence quarks. This neglects sea quarks and $q\bar{q}$ pairs which play the role of mesonic dressing. Calculations have been carried through second order in the momentum and gradients of the field. Higher order in these quantities, as well as non-linear effects in the scalar field, are model dependent. As a check on the validity of the approach, the calculations were carried out for external electromagnetic fields, where the experimental situation is clear, as well as for general isoscalar and isovector vector potentials.

The Hamiltonian is written

$$H = H_0 + H', \qquad (5.31)$$

where H_0 is the free soliton model Hamiltonian. The interaction term for isoscalar potentials is

$$H' = \tfrac{1}{3} \int d^3x\, \overline{\psi}(\vec{x}) \Big(\gamma^\mu\, V_\mu(\vec{x}) + S(\vec{x})\Big) \psi(\vec{x}). \qquad (5.32)$$

The scalar charge is chosen to be 1/3 so that the total potential acting on three quarks has a weight of unity. For the electromagnetic case

$$H'_{em} = \int d^3x \, \overline{\psi}(\vec{x}) \left(\tfrac{1}{6} + \tfrac{1}{2}\tau_3\right) \gamma^\mu \, eA_\mu(\vec{x}) \, \psi(\vec{x}). \qquad (5.33)$$

The general form for the generator coordinate wave function [compare (5.30)] including internal excitation of the bag is

$$|\Psi\rangle = \sum_n \int d^3X \left[e^{-i\vec{P}\cdot\vec{X}} e^{i\vec{K}\cdot\vec{y}} e^{i\vec{P}\cdot\vec{X}} \right] \phi_n(\vec{X}) |\vec{X}; n\rangle \qquad (5.34)$$

where n labels a complete set of orthonormal states $|\vec{X}; n\rangle$ localized about $\vec{X}$. The rapidity operator $\vec{y}$ is again replaced by the velocity operator, which, however, must be treated consistently with the effective Hamiltonian. We anticipate the final result and set

$$\vec{y} \simeq \vec{v} = \frac{1}{m} \left[-i\vec{\nabla} - e(\tfrac{1}{6} + \tfrac{1}{2}\tau_3)\vec{A} - \tfrac{1}{3}\vec{V} \right]. \qquad (5.35)$$

Remember that the $\vec{\nabla}$ operates only on ϕ_n. (We could have left the coefficients of $\vec{A}$ and $\vec{V}$ arbitrary, and then determine them consistently at the end.) Here we limit n to the magnetic quantum numbers of the nucleon, $m = \pm\tfrac{1}{2}$.

The expectation value of the energy is given by

$$\langle \Psi | H | \Psi \rangle =$$

$$\sum_{mm'} \int d^3X \int d^3X' \phi_m^*(\vec{X}) \, \langle \vec{X}; m | B^\dagger(\vec{X}) \, H \, B(\vec{X}') | \vec{X}'; m' \rangle \, \phi_{m'}(\vec{X}'),$$

$$(5.36)$$

where we have defined

$$B(\vec{X}) = e^{-i\vec{P}\cdot\vec{X}} e^{i\vec{K}\cdot\vec{v}} e^{i\vec{P}\cdot\vec{X}}. \qquad (5.37)$$

Variation of $< \Psi | H | \Psi >$ with respect to $\phi_m^*(\vec{X})$, subject to the constraint of wave function normalization, leads to a modified, coupled two-component Hill-Wheeler (1953) integral equation for the ϕ_m :

$$\sum_{m'} \int d^3X' < \vec{X}; m | B^\dagger(\vec{X}) (H - E) B(\vec{X}') | \vec{X}'; m' > \phi_{m''}(\vec{X}') = 0 .$$

(5.38)

The integral equation is replaced (see Sec. 5.2) by a differential equation obtained as an expansion in gradients of ϕ (more properly, in the velocity) and, for present purposes, in gradients of the potentials. We begin with the expectation value of $H - E$. We drop the index "m" and regard ϕ as a Pauli spinor for the nucleon. Let

$$\vec{R} = \tfrac{1}{2}(\vec{X}' + \vec{X}) , \qquad \vec{r} = \vec{X}' - \vec{X} .$$

(5.39)

Then

$$\langle \Psi | H - E | \Psi \rangle = \int d^3R \int d^3r \, \phi^\dagger(\vec{R} - \vec{r}/2)$$

$$\times \, <-\vec{r}/2 | B^\dagger_{-\vec{r}/2} (H_{-\vec{R}} - E) B_{\vec{r}/2} | \vec{r}/2 > \, \phi(\vec{R} + \vec{r}/2) ,$$

(5.40)

where

$$H_{-\vec{R}} = H_0 + \int d^3x \, \psi^\dagger(\vec{x}) \left[\left(\frac{1}{6} + \frac{\tau_3}{2} \right) e A_0(\vec{x} + \vec{R}) \right.$$

$$\left. + \tfrac{1}{3} \left(V_0(\vec{x} + \vec{R}) + \beta S(\vec{x} + \vec{R}) \right) \right] \psi(\vec{x}) .$$

(5.41)

with

$$B_{\pm \vec{r}/2} = \exp \left[i \vec{v} \cdot (\vec{K} \mp H \vec{r}/2) \right] .$$

(5.42)

We have used the translation operator $e^{\pm i \vec{P} \cdot \vec{R}}$ to translate the state vectors and the Hamiltonian by $\mp \vec{R}$. (In doing so, the translation and boost operators were commuted, giving rise to the exponential of the Hamiltonian.)

The overlap matrix element in Eq. (5.40) is a rapidly decreasing function of r; in particular for r larger than twice the radius of a bag, the matrix element is essentially zero. We therefore expand the weight functions in a Taylor series around $\vec{R}$ up to second order in derivatives:

$$\phi(\vec{R} + \vec{r}/2) = \phi(\vec{R}) + \tfrac{1}{2}r_n\partial_n\phi(\vec{R}) + \tfrac{1}{8}r_n r_m\partial_n\partial_m\phi(\vec{R}) + \cdots . \quad (5.43)$$

This is an expansion in terms of the collective momentum. The expansion is useful provided the wavelength of the collective motion, i.e. the motion of the bag as a whole, is larger than the bag radius. The external fields in H' are also expanded to second order in derivatives and the boost operator is used to the same order. One now integrates by parts to remove the gradients from $\phi^\dagger$. Variation with respect to $\phi^\dagger$ leads to a differential equation for $\phi(\vec{R})$. The coefficients of this differential equation depend on overlap matrix elements integrated over r and on the external potentials. Nonlocality in the potentials has thus been replaced by a gradient expansion. The following equation for the two-component spinor ϕ is obtained, where now $\vec{p} = \vec{\nabla}/i$:

$$\left[m + (\vec{p} - e\vec{A} - \vec{V})\frac{1}{2(m+\tilde{S})}(\vec{p} - e\vec{A} - \vec{V}) + eA_0 + V_0 + \tilde{S} \right.$$

$$- \frac{1}{2m}\vec{\Sigma}\cdot\left(\mu_p\vec{\nabla}\times e\vec{A} + \mu_{is}\vec{\nabla}\times\vec{V}\right)$$

$$+ \frac{1}{2m^2}\vec{\Sigma}\cdot\left((\mu_p - \tfrac{1}{2})\vec{\nabla}eA_0 + (\mu_{is} - \tfrac{1}{2})\vec{\nabla}V_0 - \tfrac{1}{2}\vec{\nabla}\tilde{S}\right)\times\vec{p}$$

$$\left. + \tfrac{1}{6}\langle r_c^2\rangle\nabla^2\left(eA_o + V_0 + \tilde{S}\right) \right]\phi = (E - m)\phi. \quad (5.44)$$

The quantity $\tilde{S}$ appears in the denominator of the second term, but the derivation was only carried through terms linear in $\tilde{S}$. It should be emphasized that the effective Hamiltonian is always only linear in V_0 and to a good approximation is linear in $\tilde{S}$. The literature if filled with "effective Schrödinger equations" containing terms $V_0^2 - S^2$. Such terms are not wrong as used, but are an artifact of not maintaining hermiticity in the differential equation.

The explicit forms for the various moments are

$$\langle r_c^2 \rangle = \frac{1}{m^2} \int d^3r \, \langle -\frac{\vec{r}}{2} | \left(m^2 (x^2)_{op} - m \left[\vec{K}, (\vec{x})_{op} \right]_+ + K^2 \right) | \frac{\vec{r}}{2} \rangle \quad (5.45)$$

for the mean-square size, and

$$\mu_p = \frac{2}{3} \int d^3r \, N_q^2(r) N_\sigma(r) \int d^3x \, u(\vec{x} + \vec{r}/2) \, v(\vec{x} - \vec{r}/2) \frac{x^2 - \vec{r} \cdot \vec{x}/2}{|\vec{x} - \vec{r}/2|}$$

$$\div \int d^3r \, N_q^3(r) N_\sigma(r) \quad (5.46)$$

for the proton magnetic moment. Here u and v are the upper and lower components of the quark wave functions, N_q and N_σ are single-quark and σ overlap integrals (Lübeck, et al., 1987). Here

$$\vec{x}_{op} = \int \overline{\psi}(\vec{x}) \, \vec{x} \, \psi(\vec{x}) \, d^3x , \quad (5.47)$$

and $\tilde{S} = \langle \beta \rangle S$. For the MIT bag, for example, $\langle \beta \rangle = \int d^3r (u^2 - v^2) = 0.4795$. The explicit forms for the proton charge moment $\langle r_c^2 \rangle$ and magnetic moment μ_p can be expressed in terms of the bag wave function. We assume that the model parameters have been adjusted to reproduce the experimental values: $\langle r_c^2 \rangle = (0.83 \text{ fm})^2$ and $\mu_p = 2.79285$ nuclear magnetons. The anomalous magnetic moment is $\kappa_p = \mu_p - 1 = 1.79285$.

In the case of the proton, the electric charges of the quarks add up to give a magnetic moment equal to that of one quark carrying the full charge $+e$. In the isoscalar case, each quark carries a "charge" of 1/3, so that the isoscalar "magnetic" moment is

$$\mu_{is} = \tfrac{1}{3}\mu_p \quad \text{or} \quad \kappa_{is} = (\kappa_p - 2)/3 \approx -0.07 . \quad (5.48)$$

A similar calculation for the isovector moment gives

$$\mu_{iv} = \tfrac{5}{3}\mu_p = 4.65 \quad \text{or} \quad \kappa_{iv} = 3.65 . \quad (5.49)$$

Note that $\frac{1}{2}(\kappa_{is} + \kappa_{iv}) = \kappa_p$. Although the isoscalar anomalous moment is small, the isovector anomalous moment is very large.

We can now compare these results with those obtained for a point Dirac particle (minimum coupling) in the Foldy-Wouthuysen (1950) representation.

The coefficient of the $\vec{\Sigma} \cdot \vec{B}$ term contains $(\kappa_{is} + 1)$ and the $\vec{L} \cdot \vec{\Sigma}$ term contains $\kappa_{is} + \frac{1}{2}$. Although we know of no *a priori* reason why κ_{is} should be small (see however, Bleszynski, Bleszynski and Jaroszewicz, 1987), we find (from its relationship to the electromagnetic μ_p) that it is small, in qualitative agreement with the Dirac point particle description. However, we have seen that κ_{iv} is large, and the Darwin term is larger than the Dirac value by a factor of 25. Nevertheless, it is clear that scattering phenomenology can absorb defects in the Dirac equation to the order of these calculations. The depths of the vector and scalar potentials can be adjusted to give the required depth of the central potential. The difference can be adjusted to give the required spin-orbit coupling. The nucleon size (Darwin term) can be folded into the potential forms by suitably adjusting their shapes. The utility in relativistic Lagrangian field theories is less clear, because of the crucial role played by form factors. See, for example, Cahill, Roberts and Praschifka (1987).

Chapter six

ONE GLUON EXCHANGE

To the order of one-gluon exchange, the gluon field equations linearize and are formally identical to Maxwell's equations *in media*, except that all field operators and currents are now $SU(3)$ matrices, proportional to the Gell-Mann λ-matrices.

6.1 Absolute color confinement

In order to effect color confinement, the dielectric function $\kappa(\sigma)$ is modeled to satisfy the conditions

$$\kappa(0) = 1, \qquad \kappa(\sigma_v) = 0 . \qquad (6.1)$$

This can be seen by considering, for example, a spherically symmetric system with static charge density, $Q\,\delta^3(\vec{r})$, inside a cavity with $\kappa = 1$; outside the cavity, $\kappa \to 0$ as $r \to \infty$. The chromo-electrostatic field equations are then the familiar

$$\vec{\nabla} \cdot \vec{D} = \rho \qquad\qquad \vec{\nabla} \times \vec{E} = 0$$
$$\vec{D} = \kappa \vec{E} \qquad\qquad \vec{E} = -\vec{\nabla} A_0 , \qquad (6.2)$$

where here, however, all quantities are, as already noted, color matrices. From Gauss's law we find immediately that

$$\vec{D} = \frac{Q}{4\pi r^2}\hat{r} \qquad (6.3)$$

irrespective of the functional form of $\kappa(|\vec{r}|)$, and the energy is

$$\mathcal{E} = \frac{1}{2}\int d^3r \, \vec{D} \cdot \vec{E} = \frac{1}{2}\int d^3r \, \frac{D^2}{\kappa(r)} = \frac{Q^2}{8\pi}\int_0^\infty \frac{dr}{r^2}\frac{1}{\kappa(r)} . \qquad (6.4)$$

The divergence at the lower limit of integration is associated with the usual self-energy. However, outside the cavity the integral diverges because $\kappa \to 0$ (exponentially in the model). The energy is infinite unless the charge vanishes. This conclusion does not depend on spherical symmetry. Any net charge gives rise to a $\vec{D}$-field over all space. Therefore the charge within the cavity must be in a color singlet state.

6.2 Chromo-electric confinement in an MIT cavity

We have already seen that the dielectric properties of the cavity and vacuum assure absolute color confinement: an isolated structure not in a color singlet state has infinite energy. It is instructive to expand upon this point by considering a fixed (*i.e.* massive) particle at an arbitrary point in an MIT-type cavity of radius R, with dielectric constant $\kappa = 1$ inside and $\kappa = \kappa_v$ outside. Eventually we want the limit $\kappa_v \to 0$. (Fai, Perry and Wilets, 1988.)

The static chromo-electric Green's function (in Coulomb gauge) is (see also Lee, 1979)

$$G^{00}(\vec{r}, \vec{r}') =$$

$$\frac{1}{4\pi} \sum_\ell P_\ell(\cos\theta) \left[\frac{r_<^\ell}{r_>^{\ell+1}} + \frac{(rr')^\ell}{R^{2\ell+1}} \frac{(1 - \kappa_v)(\ell+1)}{\kappa_v(\ell+1) + \ell} \right], \qquad r, r' < R.$$

(6.5)

Inserting the color charge operator $\frac{1}{2}g_s\lambda$, we find that the self energy for a quark at $\vec{r}$ is

$$E_{self}(r) = \frac{1}{2} < (\tfrac{1}{2}g_s\vec{\lambda})^2 > \lim_{\vec{r}' \to \vec{r}} G^{00}(\vec{r}, \vec{r}') = \tfrac{2}{3}g_s^2 \lim_{\vec{r}' \to \vec{r}} G^{00}(\vec{r}, \vec{r}) =$$

$$\tfrac{2}{3}\alpha_s \left\{ \frac{1}{r} \sum_\ell 1 + \frac{(1 - \kappa_v)}{R\kappa_v} + \frac{1}{R}\left[\frac{r^2}{R^2 - r^2} - \log(1 - r^2/R^2) \right] \right\}.$$

(6.6)

This is the same, of course, as the field energy. We have used $< \vec{\lambda}^2 > = 16/3$ and $\alpha_s = g_s^2/4\pi$. The first term in curly brackets in (6.6), although it may appear strange, is to be identified with the usual infinite self energy of an isolated quark in a medium of $\kappa = 1$.

The second term is a monopole ($\ell = 0$) contribution arising from the second term in (6.5). It is independent of position but becomes infinite as $\kappa_v \to 0$. The infinity is cancelled by mutual interaction with other quarks in the cavity if the total color wave function is in a singlet state. The remaining terms ($\ell \geq 1$) are finite as $\kappa_v \to 0$ and are summed to give the terms in square brackets. Note that as $r \to R$, these yield $(2\alpha_s/3)/2(R - r)$ corresponding to interaction of the quark with an image charge of the same sign.

Since the self energy adds to the mass of the quark, it can be identified here as a scalar confinement potential

$$V_{conf} = \frac{2\alpha_s}{3R} \left[\frac{r^2}{R^2 - r^2} - \ln(1 - r^2/R^2) \right]. \tag{6.7}$$

A calculation of the expectation value of V_{conf}

$$\int \bar{\psi}_0 V_{conf} \psi_0 d^3 r \,,$$

for the lowest $s_{1/2}$ quark state in the MIT bag model yields the value $1.149\alpha_s/R$. compared with the results of Goldhaber, Hansson and Jaffe (1983) who obtained $0.903\alpha_s/R$ for light quarks (see also Hansson and Jaffe, 1983). The similarity in the numerical results indicates that it may be reasonable to use the massive quark results for light quarks. This is because confinement imparts to the quarks an effective mass.

The above results can also be interpreted as the contribution to the self energy of a quark of any mass due to A_0 in the Coulomb, or transverse, gauge since then A_0 is instantaneous.

6.3 Example of confinement: the flux tube

The spectra of heavy mesons, e.g. charmonium and the Υ, have been *fit* with a non-relativistic potential of the form

$$V(r) = -\frac{4}{3} \frac{\alpha_s}{r} + \theta r + \text{const}, \tag{6.8}$$

which is consistent with the expected short range Coulomb behavior
and the long range linear confinement potential. The intermediate
range part, where the heavy meson wave functions peak, however,
has not been established definitively by lattice gauge calculations.
The coefficient of the linear term, θ, is the string constant, numerically equal to about 925 MeV/fm (see Sec. 2.3).

The string tension can be understood by considering a quark
and an antiquark separated by a large distance r (Born-Oppenheimer
approximation). Over most of the region of interest we can assume
cylindrical, axially symmetric geometry. The flux through the tube
is just equal to the charge $Q = \frac{1}{2}\lambda g_s$.

The calculation is straightforward in the MIT model. Johnson
and Thorn (1976) find for the energy per unit length

$$\theta = BA + \frac{8\pi}{3}\frac{\alpha_s}{A}, \tag{6.9}$$

where A is the cross sectional area of the cylindrical cavity, B is the
bag constant. Minimization with respect to A yields

$$A^2 = \frac{8\pi\alpha_s}{3B}, \tag{6.10}$$

hence

$$\theta = \left[\frac{32\pi}{3}\alpha_s B\right]^{1/2}. \tag{6.11}$$

Inserting the MIT values $B = 57.5$ MeV/fm^3 and $\alpha_s = 2.2$ gives close
to the experimental value of 910 MeV/fm (watch for implicit $\hbar c =$
197.33 MeV-fm). However, these parameters are the ones derived
using the dubious Casimir term (see Sec. 1.4).

The tube radius can be expressed in terms of the phenomenological parameters:

$$R_t = \sqrt{A/\pi} = \sqrt{\frac{16\alpha_s}{3\theta}} = \sqrt{\frac{\theta}{2\pi B}}.$$

For the above value of B, this yields $R_t = 1.6$ fm. This is larger than
characteristic hadron sizes. Without the Casimir term the radius
would be even larger!

Johnson (1978) has extended the heavy quark-antiquark calculations to small separations (giving up cylindrical symmetry) in
order to justify the form (6.8). Various charmonium and bottomonium calculations in the potential model using (6.) tend to put the
quark wave functions in regions where both the $1/r$ and r terms are
comparable and cannot really isolate the linear term.

The tube radius is not well determined in lattice gauge calculations, but Wosiek and Haymaker (1987) have made a study of the
energy density profile of a flux tube. Although not conclusive, they
find that the distribution perpendicular to the tube axis seems to follow an exponential fall off, with a root mean square radius $<r_\perp^2>^{1/2}$
of about 1 fm. A tube of uniform distribution of the same rms size
would have an effective radius of $R_t = \sqrt{2} <r_\perp^2>^{1/2}$. These numbers
are not inconsistent with the value of R_t quoted above.

Polarization of the vacuum by light quark-antiquark pairs leads
to a shielding of the (heavy) interquark potential, softening the confinement and eventually resulting in a breaking of the tube. (*cf.*
Vasak, Wietschorke, Müller and Greiner, 1983.) Consider the energy to create a pair of mesons, each consisting of one heavy and one
light quark say $Q\bar{q}$, $\bar{Q}q$, compared with $Q\bar{Q}$. This is about 700 MeV
for the $D\bar{D}$ threshold compared with the $J/\Psi(c\bar{c})$ mass, and about
1 GeV for the $B\bar{B}$ threshold compared with the $\Upsilon(b\bar{b})$ mass. These
correspond to a stretch of only about 1 fm, or half the diameter of
the tube. This is further indication that the region of the linear
potential is never actually attained.

The calculation in the soliton model follows similarly. The $\vec{E}$
and $\vec{D}$ fields are parallel to the symmetry axis. Since $\vec{\nabla} \times \vec{E} = 0$,
the $\vec{E}$ field is independent of $\vec{r}$, but $\vec{D}(\rho) = \kappa(\sigma(\rho))\vec{E}$, where ρ is the
distance from the symmetry axis. The energy per unit length is

$$\frac{\mathcal{E}}{L} \equiv \theta = \int \left[\tfrac{1}{2}|\vec{\nabla}\sigma|^2 + U(\sigma) \right] dA + \tfrac{1}{2}|\vec{E}|^2 \int \kappa(\sigma) dA. \qquad (6.12)$$

By Gauss's law,

$$\vec{\nabla} \cdot \vec{D} = \vec{J}_0, \qquad (6.13)$$

the flux through the tube is equal to the quark charge,

$$E \int \kappa(\sigma) dA = \tfrac{1}{2} g_s \lambda. \qquad (6.14)$$

This gives

$$\theta = \int \left[\tfrac{1}{2} |\vec{\nabla}\sigma|^2 + U(\sigma) \right] dA + \frac{g_s^2 <\vec{\lambda}\cdot\vec{\lambda}>}{8 \int \kappa(\sigma)dA}$$

$$= \int \left[\tfrac{1}{2} |\vec{\nabla}\sigma|^2 + U(\sigma) \right] dA + \frac{8\pi\alpha_s}{3 \int \kappa(\sigma)dA} . \tag{6.15}$$

Variation with respect to $\sigma(\rho)$ leads to a differential-integral equation for σ,

$$-\left(\frac{d^2}{d\rho^2} + \frac{1}{\rho}\frac{d}{d\rho} \right)\sigma(\rho) + U'(\sigma(\rho)) - \frac{2}{3\pi}\alpha_s \frac{\kappa'(\sigma(\rho))}{\left[\int \rho'd\rho'\kappa(\sigma(\rho')) \right]^2} = 0 ; \tag{6.16}$$

the boundary conditions are $\sigma'(0) = 0$ and $\sigma(\infty) = \sigma_v$. Since $U'(\sigma_v) = 0$ we find that we have another, important, condition

$$\kappa'(\sigma_v) = 0 . \tag{6.17}$$

Solutions to (6.16) for various forms of $\kappa(\sigma)$ have been presented by Bickeböller *et al* (1985), see Fig. 6.1.

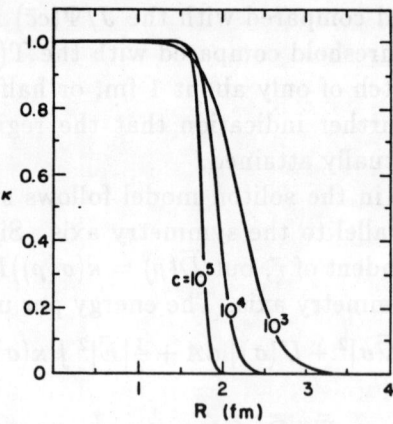

Figure 6.1. The flux tube dielectric function plotted against the perpendicular radius for three values of the model parameter c. The field energy is proportional to κ because $\vec{E}$ is constant. (Bickeböller, *et al.*, 1985).

The results shown in Fig. 6.1 are in qualitative agreement with those of Baker, Ball and Zachariesen, 1988.

The flux tube calculation is of particular interest because it does not involve quark-sigma coupling and so is independent of which form for $g(\sigma)$ is assumed. It depends only upon the parameters of $U(\sigma)$ and the form of $\kappa(\sigma)$. The chiral soliton model discussed in Chapter 8, in particular, has no direct quark-sigma coupling in the model Lagrangian although an effective coupling emerges through the quark self energy generated by the chromoelectric field.

6.4 There is no color Van der Waals problem!

The vanishing of κ in the vacuum guarantees that there is no (long-range) interaction between isolated bags. The problem of the r^{-6} [or rather r^{-7} when retardation is included, (Casimir and Polder, 1948)] Van der Waals potential plagues most potential quark models. A variety of *ad hoc* prescriptions have been proffered to eliminate the problem. Here, however, because the stress-energy tensor is bilinear in E and D, and in B and H, and therefore vanishes in the vacuum, no gluonic forces can be transmitted through the vacuum, in the Abelian approximation (Lee. 1979). (E and B can be finite in the vacuum.) Although Lee seems to have appreciated this, he nevertheless proposed a gluon mass term to prevent penetration of the vacuum by E and B. This has the undesirable features of destroying gauge invariance and absolute color confinement. At the level of the present calculations, the form assumed for κ is alone sufficient to remove the Van der Waals problem.

In some ways, the result may seem surprising, because finite $\vec{E}$ and $\vec{B}$ fields can exist in the vacuum even if $\vec{D}$ and $\vec{H}$ must vanish there. We already saw this in the case of the flux tube: $\vec{E}$ is everywhere constant, right across the flux tube surface. The matching conditions for the fields across a discontinuity in the dielectric/permittivity are that the parallel components of $\vec{E}$ and $\vec{H}$ are continuous, and the normal components of $\vec{D}$ and $\vec{B}$ are continuous. External $\vec{E}$ and $\vec{B}$ fields cannot penetrate a cavity with $\kappa \neq 0$. *Relative to the vacuum,* a cavity appears to be a perfect dielectric

$\kappa(\text{in})/\kappa(\text{vac}) \to \infty$ and a perfect diamagnet $\mu(\text{in})/\mu(\text{vac}) = 0$. Recall that everywhere $\mu\kappa = 1$. Field configurations for bags containing electric and magnetic dipoles are shown in Figs. (6.2) and (6.3). Also shown alongside are empty bags bathed in external electric and magnetic fields. The $\vec{B}$ field is expelled from an (empty) bag as it would be from a superconductor. The external $\vec{E}$ field is normal to the empty bag at the surface, but vanishes inside because normal $\vec{D}$ is continuous and vanishes outside.

6.5 The linearized gluon propagator

In order to calculate self and mutual interactions, it is necessary to first calculate the gluon propagator in medium. We follow here Bickeböller. *et al.* (1984) as corrected by Tang and Wilets (1989). In a scalar medium, the gluon propagator is diagonal in the color indices,

$$G^{cc'}_{\mu\mu'} = \delta^{cc'} G_{\mu\mu'}, \tag{6.18}$$

and we suppress the color indices in what follows.

In the linearized (one gluon exchange) approximation. the gluon propagator satisfies equations which are identical to those for a Maxwell propagator in medium. We work in the transverse gauge, so that

$$\partial_i \kappa A^i = 0 = \partial_i \kappa G^{ij}. \tag{6.19}$$

The potentials are then given by

$$A^0(\vec{r},t) = \int d^3r' \, G^{00}(\vec{r}:\vec{r}') J^0(\vec{r}',t), \tag{6.20\,a}$$

$$A^i(\vec{r},t) = \int d^3r' \, dt' \, G^{ii'}(\vec{r},t:\vec{r}',t') J_i^{i'}(\vec{r}',t'). \tag{6.20\,b}$$

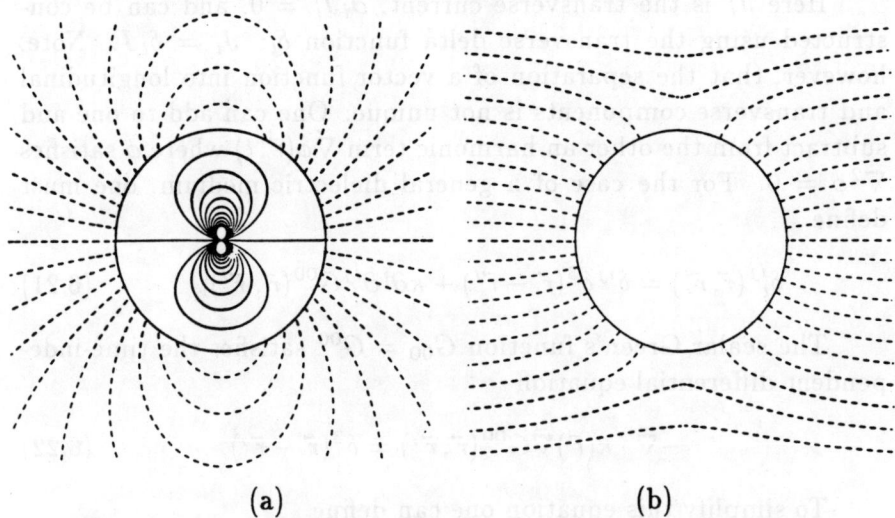

Figure 6.2. Chromo-electric fields E (dashed) and D (solid) field lines for (a) an electric dipole in an MIT cavity, and (b) for an empty cavity bathed in an external E field.

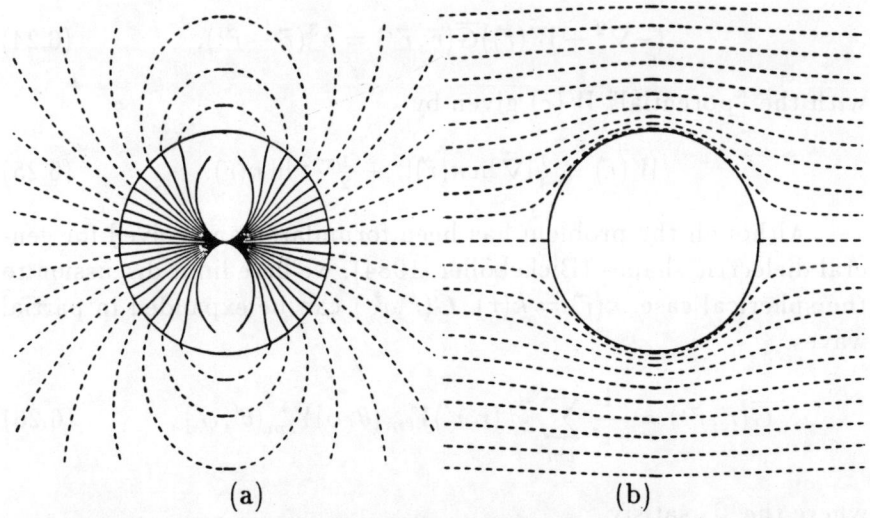

Figure 6.3. Chromo-magnetic fields B (dashed) and H (solid) field lines for (a) a magnetic dipole in an MIT cavity, and (b) for an empty cavity bathed in an external B field.

Here J_t^i is the transverse current, $\partial_i J_t^i = 0$, and can be constructed using the transverse delta function δ_t: $J_t = \delta_t J$. Note, however, that the separation of a vector function into longitudinal and transverse components is not unique: One can add to one and subtract from the other an harmonic term $\vec{\nabla}\phi(\vec{r},t)$ where ϕ satisfies $\nabla^2\phi = 0$. For the case of a general dielectric medium, one must define

$$\delta_t^{ij}(\vec{r},\vec{r}') = \delta^{ij}\delta^3(\vec{r}-\vec{r}') + \kappa\partial^i\partial'^j G^{00}(\vec{r},\vec{r}') . \qquad (6.21)$$

The scalar Green's function $G_{00} = G^{00}$ satisfies the time-independent differential equation

$$-\vec{\nabla}\cdot\kappa(\vec{r})\vec{\nabla}G^{00}(\vec{r},\vec{r}') = \delta^3(\vec{r}-\vec{r}') . \qquad (6.22)$$

To simplify this equation one can define

$$\overline{G}(\vec{r},\vec{r}') \equiv \sqrt{\kappa(\vec{r})}\, G(\vec{r},\vec{r}')\sqrt{\kappa(\vec{r}')}, \qquad (6.23)$$

where $\overline{G}$ satisfies

$$(-\nabla^2 + W(\vec{r}))\overline{G}(\vec{r},\vec{r}') = \delta^3(\vec{r}-\vec{r}') . \qquad (6.24)$$

with the "potential" $W(\vec{r})$ given by

$$W(\vec{r}) \equiv \tfrac{1}{4}|\vec{\nabla}\ln\kappa(\vec{r})|^2 + \tfrac{1}{2}\nabla^2\ln\kappa(\vec{r}) . \qquad (6.25)$$

Although the problem has been formulated and solved for general dielectric shapes (Bickeböller. 1984), we here limit discussion to the spherical case. $\kappa(\vec{r}) = \kappa(r)$. $\overline{G}(\vec{r},\vec{r}')$ can be expanded in partial waves.

$$\overline{G}(\vec{r},\vec{r}') = \frac{1}{rr'}\sum_{\ell m}\overline{\mathcal{G}}_\ell(r,r')Y_{\ell m}(\theta.\phi)Y_{\ell m}^*(\theta',\phi') , \qquad (6.26)$$

where the $\overline{\mathcal{G}}_\ell$ satisfy

$$\left[-\frac{d^2}{dr^2} + \frac{\ell(\ell+1)}{r^2} + W(r)\right]\overline{\mathcal{G}}_\ell(r,r') = \delta(r-r') . \qquad (6.27)$$

Let $f_\ell(r)$ and $g_\ell(r)$ be solutions of the homogeneous equation

$$\left[-\frac{d^2}{dr^2} + \frac{\ell(\ell+1)}{r^2} + W(r)\right]\left\{\begin{matrix} f_\ell(r) \\ g_\ell(r) \end{matrix}\right\} = 0 \qquad (6.28)$$

which are regular at $r = 0$ and $r \to \infty$, respectively. Then

$$\overline{\mathcal{G}}_\ell(r, r') = -\frac{f_\ell(r_<)g_\ell(r_>)}{w\{f_\ell, g_\ell\}}, \qquad (6.29)$$

with the Wronskian $w\{f, g\} = fg' - f'g$. For the case $\kappa = 1$, the above reduces to the Coulomb result $\overline{\mathcal{G}}_\ell = (2\ell + 1)^{-1} r_<^{\ell+1}/r_>^\ell$ and $G = 1/4\pi|\vec{r} - \vec{r}'|$.

The differential equation for the vector potential is time-dependent. After a Fourier transformation in the time and defining the transverse dyadic Green's function as

$$\overline{G}^{ii'}(\vec{r}.\vec{r};\omega) \equiv \kappa(\vec{r})G^{ii'}(\vec{r}.\vec{r}'.\omega). \qquad (6.30)$$

one obtains

$$(\omega^2 + \nabla^2)\overline{G}^{ii'}(\vec{r}, \vec{r}', \omega) - \epsilon_{ikl}\partial^k\left(\epsilon_{lmn}\overline{G}^{mi'}(\vec{r}.\vec{r}'\omega)\partial^n \ln \kappa(\vec{r})\right)$$
$$= -\delta_t^{ii'}(\vec{r}.\vec{r}'). \qquad (6.31)$$

with the restriction $\partial^i\overline{G}^{ii'}(\vec{r}.\vec{r}'.\omega) = 0$. The transverse gauge condition requires the asymmetric definition (6.30) of $\overline{G}$.

For the free case. $\kappa = 1$. the transverse delta-function assumes the form

$$\delta_{t,f}^{ii'}(\vec{r} - \vec{r}') = \frac{\delta(r - r')}{r^2} \left\{ \mathcal{Y}_{llm}^i(\Omega) \mathcal{Y}_{llm}^{i'*}(\Omega') \right.$$

$$+ \left(\sqrt{\frac{l}{2l+1}} \mathcal{Y}_{l,l+1,m}^i(\Omega) + \sqrt{\frac{l+1}{2l+1}} \mathcal{Y}_{l,l-1,m}^i(\Omega) \right)$$

$$\times \left(\sqrt{\frac{l}{2l+1}} \mathcal{Y}_{l,l+1,m}^{i*}(\Omega') + \sqrt{\frac{l+1}{2l+1}} \mathcal{Y}_{l,l-1,m}^{i'*}(\Omega') \right) \right\}$$

$$- \sqrt{l(l+1)} \left\{ \theta(r - r') \frac{r'^{l-1}}{r^{l+2}} \mathcal{Y}_{l,l+1,m}^i(\Omega) \mathcal{Y}^{i'*}{}_{l,l-1,m}(\Omega') \right.$$

$$\left. + \theta(r' - r) \frac{r^{l-1}}{r'^{l+2}} \mathcal{Y}_{l,l-1,m}^i(\Omega) \mathcal{Y}^{i'*}{}_{l,l+1,m}(\Omega') \right\}. \tag{6.32}$$

The $\vec{\mathcal{Y}}_{ljm}$ are vector spherical harmonics, and are not to be confused with the spinor harmonics $\mathcal{Y}_{jm}^\ell$ introduced earlier. In particular, $\vec{\mathcal{Y}}_{jjm} = [j(j+1)]^{-1/2} \vec{L} Y_{jm}$. Note that $\delta_{t,f}$ contains pieces with power law behavior.

An important property of the free transverse delta-function, not possessed by the general one, is that it is an identity operator when acting on any transverse vector function:

$$a^i(\vec{r}) = \int d^3r' \delta_{t,f}^{ii'}(\vec{r}, \vec{r}') a^{i'}(\vec{r}'). \tag{6.33}$$

Bickerböller *et al.* displayed the Green's functions obtained by solving the differential equation with general κ but using the free transverse delta function for the right hand side. It follows from (6.33) that one can use their results if one acts with the Green's function on the proper transverse current, $J_t = \delta_t J$ with δ_t given by (6.21).

The form of the general solution can be found in Bickeböller *et al.* We present here results for some cases of special interest.

Of particular interest is the M1 mode which, for $\omega = 0$, is

$$G_{M1}^{ii'}(\vec{r}, \vec{r}', 0) =$$

$$\left[\frac{\kappa(r_0)}{w(r_0)}\right]\frac{\tilde{f}_1(r_<)\tilde{g}_1(r_>)}{r\kappa(r)r'\kappa(r')}\sum_m\left[\vec{\mathcal{Y}}_{11m}(\Omega_<)\right]^{i_<}\left[\vec{\mathcal{Y}}_{11m}^*(\Omega_<)\right]^{i_<}, \quad (6.34\,a)$$

where

$$\left[-\frac{d^2}{dr^2}+\frac{2}{r^2}-\left(\frac{d}{dr}\ln\kappa\right)\frac{d}{dr}+\left(\frac{d^2}{dr^2}\ln\kappa\right)\right]\left\{\begin{array}{c}\tilde{f}_1(r)\\\tilde{g}_1(r)\end{array}\right\}=0,$$

$$(6.34\,b)$$

with f_1 regular at the origin and g_1 regular at infinity; r_0 is an arbitrary point where the Wronskian w is evaluated. Coordinates or indices carrying the subscript $_<$ or $_>$ are to be identified with the corresponding $r_<$ or $r_<$.

For the case $\kappa = 1$, but $\omega \neq 0$, Bickeböller *et al.* give

$$G^{ii'}(\vec{r},\vec{r}',\omega)=-\omega[j_l(\omega r_<)\vec{\mathcal{Y}}_{llm}(\Omega_<)]^{i_<}[n_l(\omega r_>)\vec{\mathcal{Y}}_{llm}^*(\Omega_>)]^{i_>}$$
$$-\frac{1}{\omega}[\vec{\nabla}\times j_l(\omega r_<)\vec{\mathcal{Y}}_{llm}(\Omega_<)]^{i_<}[\vec{\nabla}\times n_l(\omega r_>)\vec{\mathcal{Y}}_{llm}^*(\Omega_>)]^{i_>}$$
$$-\frac{1}{\omega^2}\frac{1}{2l+1}[\vec{\nabla}\times r_<^l\vec{\mathcal{Y}}_{llm}(\Omega_<)]^{i_<}[\vec{\nabla}\times\frac{1}{r_>^{l+1}}\vec{\mathcal{Y}}_{llm}^*(\Omega_>)]^{i_>},$$

$$(6.35)$$

where j_l and n_l are the spherical Bessel functions. The first term has the magnetic (TE) modes and the second and third term compose the electric (TM) modes. The non-local parts of the transverse delta function contribute only to the third term.

The static limit, $\omega \to 0$, is

$$G^{ii'}(\vec{r},\vec{r}',0)=\frac{1}{2l+1}\frac{r_<^l}{r_>^{l+1}}[\vec{\mathcal{Y}}_{llm}(\Omega_<)]^{i_<}[\vec{\mathcal{Y}}_{llm}^*(\Omega_>)]^{i_>}$$
$$-\frac{1}{2(2l+1)(2l+3)}\left(\vec{\nabla}\times r_<^{l+2}\vec{\mathcal{Y}}_{llm}(\Omega_<)\right)^{i_<}\left(\vec{\nabla}\times\frac{1}{r_>^{l+1}}\vec{\mathcal{Y}}_{llm}^*(\Omega_>)\right)^{i_>}$$
$$-\frac{1}{2(2l+1)(1-2l)}\left(\vec{\nabla}\times r_<^l\vec{\mathcal{Y}}_{llm}(\Omega_<)\right)^{i_<}\left(\vec{\nabla}\times\frac{1}{r_>^{l-1}}\vec{\mathcal{Y}}_{llm}^*(\Omega_>)\right)^{i_>}.$$

$$(6.36)$$

It is surprising to note that the Maxwell propagator in terms of spherical harmonics for a *homogeneous* medium ($\kappa = 1$) had not been formulated correctly until 1979, when it was presented by Johnson, Howard and Dudley (1979). The essential stumbling block had been the failure to realize that the transverse delta function is non-local!

The MIT bag boundary conditions are that $\hat{r} \cdot \vec{E}$ and $\hat{r} \times \vec{B}$ vanish at the bag surface, $r = R$. Since $\kappa = 1$ inside the bag, we can use the free results of Eq. (6.35) with modification. To the magnetic term we add a solution of the homogeneous equation such that the derivative of r times the function with respect to $r_>$ vanishes at the bag radius:

$$n_l(\omega r) \to n_l(\omega r) - j_l(\omega r)[R\, n_l(\omega R)]'/[R\, j_l(\omega R)]'. \qquad (6.37\,a)$$

Similarly, in the second term of (6.35), the radial function of $r_>$ must vanish at the boundary:

$$n_l(\omega r) \to n_l(\omega r) - j_l(\omega r)n_l(\omega R)/j_l(\omega R). \qquad (6.37\,b)$$

Finally, the transverse delta function requires modification. This is done by noting that one can add to $1/|\vec{r} - \vec{r}'|$ in the definition any function which satisfies $\nabla^2 f = 0$. This leads to the replacement, in the third term on the RHS of (6.35),

$$\frac{1}{r^{l+1}} \to \frac{1}{r^{l+1}} - \frac{r^l}{R^{2l+1}}. \qquad (6.37\,c)$$

For many purposes, such as spherical, static bag states, it is often sufficient to use a free propagator with an effective α_s (which is larger than for the confined case). However, when dealing with deformed bags, such as in NN collisions, the confined propagator depends upon shape, as noted by DeTar (1978). Furthermore, one must also include the chromo-electric self energy in order to assure color confinement. This is discussed further in Chapter 8.

6.6 One gluon exchange hyperfine structure

Consider the case of a bag with all quarks in the lowest $(s_{1/2})$ mean field states. The OGE contribution to the chromomagnetic hyperfine structure is given in terms of the tensor Green's function, and can be written

$$E_M = -\sum_{ckk'}\sum_{i<j}\int d^3r' J_i^{kc}(\vec{r})G^{kk'}(\vec{r},\vec{r}',0)J_i^{k'c}(\vec{r}').\qquad(6.38)$$

The frequency occurring in G has been set equal to zero since the quarks are in the same spatial state. Using the representation for ψ given by (3.9), the current for the ith quark can be expressed as

$$\vec{J}_i^c(\vec{r}) = -g_s u(r)v(r)\chi_i^\dagger \lambda^c \hat{r} \times \vec{\sigma}\chi_i.\qquad(6.39)$$

For $s_{1/2}$ quarks, only the M1 gluonic mode contributes. Then the expectation value of the magnetic energy is

$$E_M = -\tfrac{2}{3}\alpha_s \sum_{i<j} <\vec{\lambda}_i \cdot \vec{\lambda}_j\,\vec{\sigma}_i \cdot \vec{\sigma}_j> \mathcal{M},\qquad(6.40)\,a)$$

$$\mathcal{M} = \frac{\kappa(r_0)}{w(r_0)}\int r^2 dr \int r'^2 dr' \left[\frac{u(r)v(r)}{r\kappa(r)}f_1(r_<)g_1(r_>)\frac{u(r')v(r')}{r'\kappa(r')}\right].$$
$$(6.40\,b)$$

Note that the u and v are normalized according to Eq. (3.12 d), which differs from Bickebóller *et al.* by a factor of 4π.

We can use the fact that the color functions factorize here. For baryons, the color matrix elements are

$$<\vec{\lambda}_1 \cdot \vec{\lambda}_2> = -8/3\qquad(6.41)$$

and the spin matrix elements are

$$<\vec{\sigma}_1 \cdot \vec{\sigma}_2> = \begin{cases} -1 & \text{for the nucleon,} \\ +1 & \text{for the }\Delta. \end{cases}\qquad(6.42)$$

Thus the nucleon-Δ splitting is

$$E(\Delta) - E(N) = \tfrac{32}{3}\alpha_s \mathcal{M}(\text{baryon}) . \qquad (6.43)$$

Note that the Δ and the N split symmetrically about the mean value, to lowest order in α_s.

For mesons, the corresponding color and spin matrix elements are

$$<\vec{\lambda}_1 \cdot \vec{\lambda}_2> = -16/3 \qquad (6.44)$$

and

$$<\vec{\sigma}_1 \cdot \vec{\sigma}_2> = \begin{cases} -3 & \text{for the pion,} \\ +1 & \text{for the } \rho \text{ or } \omega. \end{cases} \qquad (6.45)$$

Since there is just one $q\bar{q}$ pair in a meson and three qq pairs in a baryon, the overall ρ/ω-π coefficient for the spitting is $4/3$ that for the N-Δ:

$$E(\rho) - E(\pi) = \tfrac{128}{9}\alpha_s \mathcal{M}(\text{meson}) . \qquad (6.46)$$

Of particular note is that the pion is strongly lowered in energy.

Chapter seven

PARAMETER SETS AND RESULTS

7.1 Parameterizing the model

There are five parameters in the Friedberg-Lee models, a, b, c, g_0 and α_s. Of these, c g_0 and α_s are dimensionless; $[b] = L^{-1}$ and $[a] = L^{-2}$. It is convenient to use the dimensionless "family" parameter $f = b^2/ac$ as a fourth dimensionless parameter. This leaves one dimensioned parameter (say b) to scale all lengths. All calculations have been scaled to fit the proton size $<r_p^2>^{1/2} = 0.83$ fm and parameters adjusted to fit the nucleon (or baryon) mass. This leaves three parameters available to survey, fit and predict other physical data.

Since the model is an effective one, incorporating into the parameters the contributions of neglected higher order effects, the "best fit" parameters are dependent on the level of approximation employed.

Although the parameters may vary widely with the level of approximation, the predictions of physical quantities are more stable, and, in general, improve with improvement in the approximations.

7.2 Classical mean field approximation

The classical MFA without gluonic or recoil corrections has been studied by Goldflam and Wilets (1982); Saly (1983); Köppel and Harvey (1985); Dodd and Lohe (1985).

Horn, *et al.* (1985) incorporated certain approximate recoil corrections into an extensive survey of parameters. Consistent with treating the σ-field classically, the recoil corrections only included quarks: The mass was calculated by subtracting the quark center-of-mass energy,

$$m = \sqrt{<H>^2 - <P_q^2>} \,. \qquad (7.1)$$

This was identified with the mean baryon mass $\overline{m} = \frac{1}{2}(m_N + m_\Delta) = 1087$ MeV $= 5.509$ fm^{-1}. Recoil corrections were also applied to the baryon size according to the formula of Dethier *et al.* (1983),

$$<r^2> = \left[1 - \frac{2\,\epsilon}{<H>} + \frac{A\,\epsilon^2}{<H>^2} \right] <r^2>_0 + \frac{3\,(A-1)}{4<H>^2} \,, \qquad (7.2)$$

where for baryons, $A = 3$; ϵ is the quark eigenvalue. The results were scaled to fit $<r_p^2>$. Since $\alpha_s = 0$, this left two parameters, which were taken to be $f = b^2/ac$ and c. Incidentally, the glueball mass, $m_{GB} = [U''(\sigma_v)]^{1/2}$ depends primarily on c. Selected results are presented in Table 7.1 and graphically in Fig. 7.1.

Two key pieces of data, μ_p and g_A/g_V cannot be fit (at this level) by searching on f and c, although one can obtain significant improvements over the MIT and SLAC bags. In general, μ_p would prefer a small c and g_A/g_V would prefer a large c.

The most systematic fitting of "key" data has been presented in Lübeck *et al.* (1987) The calculations go up to the level of including:

1. Momentum projection using the generator coordinate method. The σ-field was treated quantally in the generalized coherent state (single mode) approximation. The energy was optimized by applying scale factors to the quark and sigma field mean field solutions. The σ mode frequency was also varied. (see Chapters 3 and 4).

2. One gluon exchange was included, but the gluon propagator was treated as free ($\kappa = 1$).

3. The dressing of the baryons by mesons (*i.e.* pions) was not included here, but was considered by Dethier *et al.*

Bickeböller (1986), and Dodd and Williams (1988) have included OGE terms in to the calculations self-consistently.

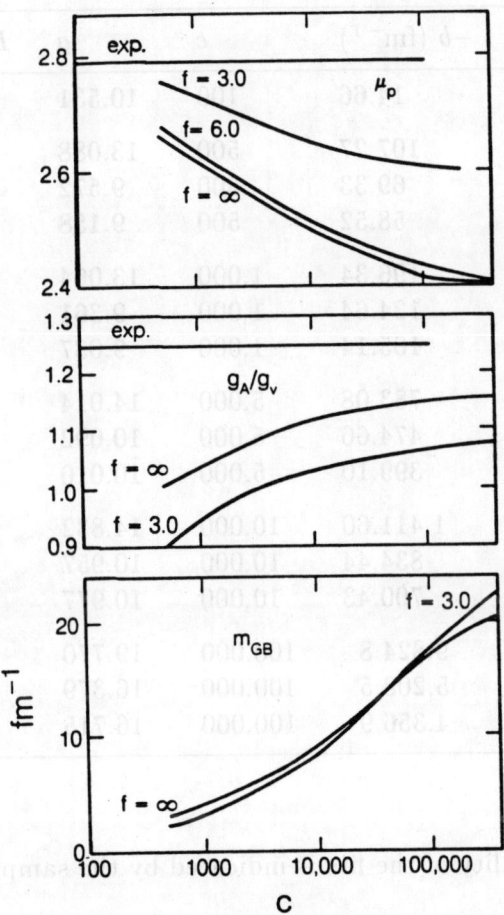

Figure 7.1. Dependence of the various quantities on the model parameter c, for two families, $f = b^2/ac$.

Table 7.1. Parameter sets which yield the same recoil corrected proton rms radius [Eq. (7.2)] of 0.83 fm and recoil corrected proton mass [Eq. (7.1)] of 939 Mev.

$a(\text{fm}^{-2})$	$-b\ (\text{fm}^{-1})$	c	g	$B\ (\text{MeV/fm}^3)$
0.000	14.66	100	10.531	10.25
7.671	107.27	500	13.088	0.00
1.602	69.33	500	9.572	16.85
0.000	58.52	500	9.158	20.83
12.849	196.34	1,000	13.064	0.00
2.589	124.64	1,000	9.361	22.01
0.000	105.14	1,000	9.037	27.12
40.880	783.08	5,000	14.014	0.00
7.510	474.66	5,000	10.092	37.03
0.000	399.10	5,000	10.010	45.06
66.422	1,411.60	10,000	14.832	0.00
11.605	834.44	10,000	10.957	44.21
0.000	700.43	10,000	10.977	53.43
321.750	9,824.8	100,000	19.770	0.00
45.214	5,208.5	100,000	16.379	67.11
0.000	4,356.9	100,000	16.715	80.07

The quality of the fits is indicated by the sampling displayed in Table 7.2.

The Δ-N splitting is too small by about half. The additional splitting can be accounted for by explicit coupling of pions (see the following section). The pion mass is reduced dramatically from its mean field, no gluon exchange value. In fact, a small increase in α_s can reduce it to the experimental value or even to zero. This should not be construed to mean that the $q\bar{q}$ description of the pion is complete. It is a much more complex structure.

The value of α_s in this fit is large, 2.25, but use of a gluon propagator confined by the bag's dielectric function leads to a coupling constant of roughly half this values.

Table 7.2. Nucleon and meson properties for the least square fit parameters with $\alpha_s = 2.25$, $g = 5.694$ and $b^2/ac = 5.438$. For each column, the parameters a and b are adjusted to give the charge rms radius for the proton: 0.83 fm. The final parameter set, "Proj. + Var." is $a = 1.908$ fm^{-2}, $b = -42.945$ fm^{-1}. $c = 177.734$, which yield $m_{GB} = 609$ MeV (rather soft), $\sigma_v = 0.621$ fm^{-1} and $B = 48.3$ MeV/fm^3. Magnetic moments are given in the ratio to the self-consistent magneton, $(2E_N)^{-1}$ (as calculated).

		MFA	Proj.	Proj.+Var.	Expt.
E_N	(MeV)	1218	1031	930	939
$<r_p^2>^{1/2}$	(fm)	0.83	0.83	0.83	0.83
μ_p	—	2.06	2.96	2.67	2.76
E_Δ	(MeV)	1384	1177	1083	1232
$<r_\Delta^2>^{1/2}$	(fm)	0.83	—	1.20	—
μ_Δ	—	—	—	3.21	—
E_π	(MeV)	950	704	177	140
$<r_\pi^2>^{1/2}$	(fm)	0.71	0.70	(~ 0)	0.66
f_π/m_π	—	—	0.04	0.83	0.66
$E_{\rho,\omega}$	(MeV)	1103	840	723	770
$<r_{\rho,\omega}^2>^{1/2}$	(fm)	—	—	1.40	—

7.3 The pion and dressing of the baryons

Dethier *et al.* (1985) have described the pion as a $q\bar{q}$ bag, and this structure has been coupled to nucleons and deltas using the coupled-state generator coordinate method, including one gluon exchange, see Figs. 7.2. DeGrand *et al.* (1975) had noted that for the MIT static bag, OGE splits the pion from the rho and omega mesons, bringing it to within a factor of two of the physical pion mass for the same parameters that fit the nucleon and the delta. This in itself was an encouraging result. Removal of spurious center-of-mass energy by projection leads to a further reduction in the pion mass. As we have seen above, full variation of the projected pion state vector comes close to yielding the physical pion mass. A 17% increase in α_s with respect to the overall best fit value yields the observed pion mass.

Williams and Dodd (1988) introduced the pion as an elementary particle in the soliton bag model, in analogy with the cloudy bag model, in order to fulfill PCAC.

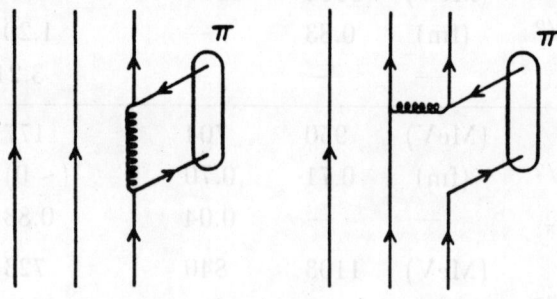

Figure 7.2. Gluon exchange diagrams coupling baryon and pion bags. (Dethier *et al.*, 1985)

In Dethier's calculation, parameters of the model are adjusted so that the composite structure reproduces the nucleon mass, the proton charge radius and the N-Δ mass-splitting. The coupling of the pion bag to the nucleon and delta permits a fitting of the delta-nucleon splitting, and is responsible for roughly 40% of that splitting. The resultant picture is thus complete and rather satisfactory.

The neutron charge radius is calculated to be $<r^2>_n= -0.08$ fm^2 compared with the experimental value of $-0.12\ \text{fm}^2$. The resulting description is similar to the Cloudy Bag Model. The probability of a "pion" in a nucleon is calculated to be 47%, indicating that two pion channels may be important. An interesting feature of the model is the partial restoration of PCAC in an integral sense: The radial integral of the divergence of the axial current is $-0.12\ \text{fm}^{-1}$ compared with 0.35 for the q^3 nucleon core and 1.25 for the MIT model. The effective "pion" tail is too short (see Fig. 7.3); this is characteristic of Hartree-type calculations, since the "eigenvalue" associated with the "pion" is lower than the separation energy (which should be the pion mass); the difference is the "rearrangement energy." The nucleon-pion coupling constant is *calculated* to be 0.19 compared with the experimental value of 0.28.

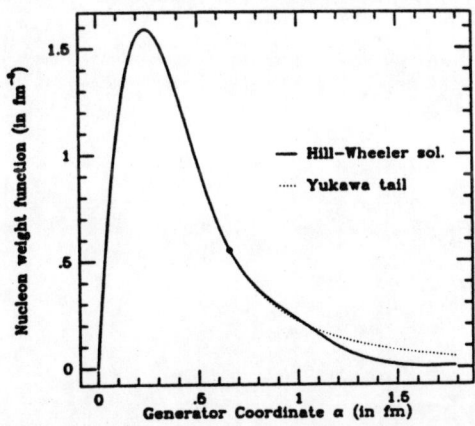

Figure 7.3. The weight function for a pion bag coupled to a three quark nucleon bag. (Dethier *et al.*, 1985)

Chapter eight

A CHIRALLY-INVARIANT
CHROMO-DIELECTRIC SOLITON MODEL

8.1 Gluon confinement

We have seen that the dielectric properties of the medium lead
to absolute confinement, *i.e.* the chromo-electric field energy is infinite for a color non-singlet cluster. This was the original motivation
for the Friedberg-Lee model, and has been pursued by a number
of authors: Nielsen and Patkos (1982), Chanfray, Nachtmann and
Pirner (1984), Pirner, Wroldsen and Ilgenfritz (1987), Bayer, Forkel
and Weise (1986).

The nontopological soliton has been studied extensively for the
case where 'confinement' has been effected by coupling of the quarks
to the sigma field. It is the quark-sigma coupling which breaks chiral
invariance: in a model without elementary pions it leads to a nonvanishing of the divergence of the axial current.

Fai, Perry and Fai (1988) proposed a variation of the Friedberg-
Lee model in which one can effect chiral invariance by removing the
direct quark-sigma coupling and provide confinement through the
chromo-electric self interaction. The energy of the gluon field due to
a single quark (as calculated in the previous chapter) can, of course,
be identified as the quark self energy. We now follow the approach
of Fai, *et al*.

8.2 More on the chromoelectrostatic self energy of a quark

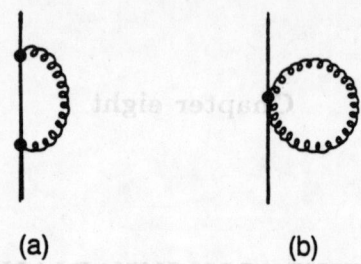

(a)   (b)

Figure 8.1. Diagram of the quark self energy in the one loop approximation for (a) finite mass quarks and (b) infinite mass quarks.

The quark self energy diagram is depicted in Fig. 8.1 a. For a fixed (infinitely massive) quark, the quark propagator shrinks to a delta function, Fig. 8.1 b. As noted in Sec. 6.3, in the Coulomb or transverse gauge, this is also the contribution of A_0 to quarks of arbitrary mass since A_0 is instantaneous. In the Abelian approximation, the self energy is given by

$$E_{self} = \tfrac{2}{3} g_s^2 \lim_{\vec{r}' \to \vec{r}} G^{00}(\vec{r}, \vec{r}').$$ (8.1)

There are two infinities associated with (8.1):

The infrared, or long range, divergence is associated with color confinement. For quarks in a color singlet state within a cavity, the mutual interaction terms cancel the (positive) divergent self energy terms. In the case of a spherical cavity, this occurs in the $\ell = 0$ electric mode. For deformed cavities, the cancellation occurs over several partial waves. Numerical calculations thus require an infrared regularization. This can be effected by the introduction of a small lower limit to the vacuum value of κ; call it κ_v. The limit $\kappa_v \to 0$ is well defined. This permits the calculation, for example, of fissioning bags where quarks confined to separated sections must be in singlet states.

There is also an ultraviolet, or short range, divergence which is weak or absent in QCD because of asymptotic freedom. It is not sufficient to subtract the free ($\kappa = 1$) self energy because the divergence is spatially dependent and cannot be removed by renormalization of the free quark mass.

In order to regularize the short distance behavior, a form factor $f(|\vec{r} - \vec{r}'|)$ is introduced; it is normalized to $\int d^3r f(r) = 1$. For example, if we choose

$$f(r) = \frac{e^{-r^2/a^2}}{a^3 \pi^{3/2}}, \tag{8.2}$$

then the divergent term will be shown to be

$$\tfrac{2}{3}\alpha_s \frac{1}{\kappa} \int d^3r f(r)/r = \frac{4\alpha_s}{3\sqrt{\pi}a\,\kappa}. \tag{8.3}$$

The divergence is proportional to a^{-1} as $a \to 0$ and would be linear in a momentum cut-off. As in QED, if the quark is described by a finite mass Dirac propagator, the divergence is logarithmic. In either case, the self energy is dominated by a term proportional to $1/\kappa$ which must be regulated. We now show this in detail for the case of spherical symmetry using the Gaussian form factor (8.2).

We consider

$$E_{self}(r) = \tfrac{2}{3}g_s^2 \int d^3r' G(\vec{r}, \vec{r}') f(|\vec{r} - \vec{r}'|) \tag{8.4}$$

where $G(\vec{r}, \vec{r}') \equiv G^{00}(\vec{r}, \vec{r}')$ is given by (6.23)-(6.28) and f is given by (8.2). Then

$$E_{self}(r) = \tfrac{2}{3}\alpha_s \sum_\ell (2\ell + 1) \int_0^\infty \frac{r'^2 dr'}{r\,r'} \frac{1}{\sqrt{\kappa(r)\kappa(r')}} \frac{f_\ell(r_<)g_\ell(r_>)}{w(g_\ell, f_\ell)}$$

$$\times 2\pi \int_{-1}^1 d\cos\widehat{rr'} \frac{e^{-|\vec{r} - \vec{r}'|^2/a^2} P_\ell(\cos\widehat{rr'})}{a^3 \pi^{3/2}}. \tag{8.5}$$

We evaluate

$$\frac{e^{-(r^2+r'^2)/a^2}}{a^3 \pi^{3/2}} 2\pi \int_{-1}^1 dx e^{2rr'x/a^2} P_\ell(x) = \frac{4}{a^3 \pi^{1/2}} e^{-(r^2+r'^2)/a^2} i_\ell(2rr'/a^2) \tag{8.6}$$

where $i_\ell(z)$ is the modified spherical Bessel function of the first kind (*cf.* Abramowitz and Stegun, 1965, p. 469):

$$i_\ell(z) \equiv \sqrt{\frac{\pi}{2z}} I_{\ell+1/2}(z) = i^{-\ell} j_\ell(i\,z)\,, \qquad (8.7)$$

with the limiting forms

$$i_\ell(z) \rightarrow \begin{cases} z^\ell/(2\ell+1)!! & \text{for } z \rightarrow 0, \\ e^z/2z & \text{for } z \rightarrow \infty\,. \end{cases} \qquad (8.8)$$

Then

$$E_{self}(r) = \tfrac{2}{3}\alpha_s \frac{4}{a^3 \pi^{1/2} \sqrt{\kappa(r)}} \sum_\ell \frac{(2\ell+1)}{w(g_\ell, f_\ell)}$$

$$\times \int \frac{r'dr'}{r\sqrt{\kappa(r')}} f_\ell(r_<)g_\ell(r_>)\, e^{-(r^2+r'^2)/a^2} i_\ell(2rr'/a^2)\,. \qquad (8.9)$$

To obtain a qualitative understanding of the behavior of the effective potential, we set

$$2z\,e^{-z}i_\ell(z) \approx \theta(\ell_c - \ell)\,, \qquad (8.10)$$

where

$$\ell_c = \xi\sqrt{z} \qquad (8.11)$$

with ξ a number of the order of unity. Then

$$E_{eff}(r) \approx$$

$$\tfrac{2}{3}\alpha_s \frac{1}{r^2\sqrt{\kappa(r)}} \sum_{\ell=0}^{\xi\sqrt{2rr'/a^2}} \frac{(2\ell+1)}{w(g_\ell, f_\ell)} \int dr' \frac{f_\ell(r_<)g_\ell(r_>)}{\sqrt{\kappa(r')}} \frac{e^{-(r-r')^2/a^2}}{\pi^{1/2}a}\,. \qquad (8.12)$$

For small a we can treat the Gaussian factor as a δ-function:

$$E_{eff}(r) \approx \tfrac{2}{3}\alpha_s \frac{1}{r\kappa(r)} \sum_{\ell=0}^{\xi\sqrt{2r/a}} \frac{(2\ell+1)}{w(g_\ell, f_\ell)} f_\ell(r)g_\ell(r)\,. \qquad (8.13)$$

Finally, since the cut-off in ℓ is very large for small a, we can use the asymptotic forms for f_ℓ and g_ℓ, which give

$$\frac{f_\ell(r)\, g_\ell(r)}{w(g_\ell, f_\ell)} \rightarrow \frac{r}{(2\ell + 1)}, \tag{8.14}$$

so that we obtain

$$E_{eff}(r) \approx \tfrac{2}{3}\alpha_s \frac{\xi\sqrt{2}}{a\,\kappa(r)}. \tag{8.15}$$

Comparison with Eq. (8.3) (justified below) suggests that $\xi = \sqrt{2/\pi}$.

The above is the basis for taking the effective quark-sigma coupling to be given by

$$g_{eff}(\sigma) = g_0\sigma_v \left[\frac{1}{\kappa(\vec{r})} - 1\right] \tag{8.16}$$

with the identification

$$g_0\sigma_v = \frac{4\alpha_s}{3\sqrt{\pi}a}. \tag{8.17}$$

The term -1 in brackets effects the quark mass renormalization: $g(\sigma)$ vanishes for $\sigma = 0$, where $\kappa = 1$. The quantity σ_v has been introduced in (8.16) to display the dimensionality of the effective coupling and to facilitate comparison with other work. Only the product $g_0\sigma_v$ appears.

We present here numerical calculations of $E_{self}(r)$, as given by Eq. (8.9), using an assumed chromo-dielectric function

$$\kappa(r) = \frac{1}{1 + e^{(r-R)/s}}, \tag{8.18}$$

for $R = 1.0$ fm and $s = 0.3$ fm. The results are displayed in Fig. 8.2 for three values of the form factor width a (with the value at the origin subtracted). The asymptotic ($a \rightarrow 0$) effective coupling (8.16) is also shown, as is $\kappa(r)$ (right scale). It can be seen that the confinement potential rises more rapidly with decreasing value of the width a, and the results of the form-factor calculation approach the result of the δ-function approximation, g_{eff}.

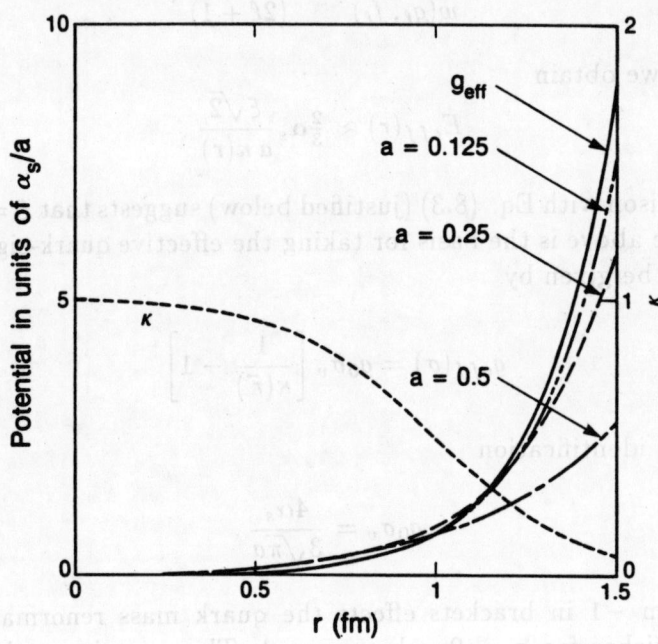

Figure 8.2. The regularized self energy (with the value at the center subtracted) for several choices of the width of the form factor, and the asymptotic effective coupling (solid line). The chromo-dielectric function $\kappa(r)$ is also shown (short dashes, right scale). [The self-energy for $a = 0.125$ fm was obtained by translating the result calculated for $R = 0.5$ fm.] (From Fai, Perry and Wilets, 1988.)

The effective self-consistent nucleon (3 quarks in the same spatial state ψ_0) mean field equations can be summarized by the following coupled nonlinear integro-differential equations:

$$\left[\vec{\alpha} \cdot \vec{p} + \beta \left(m + E_{self}(r) - \frac{4\alpha_s}{3\sqrt{\pi}a} \right) + V_x(r) - \epsilon_0 \right] \psi_0(\vec{r}) = 0,$$
(8.19)

$$-\nabla^2 \sigma + U'(\sigma) + 6g_s^2 \kappa'(\sigma) \left[\int \vec{\nabla} G(\vec{r}, \vec{r}\,') |\psi_0(\vec{r}\,')|^2 d^3r\,' \right]^2 = 0, \quad (8.20)$$

where E_{self} is defined in Eq. (8.9) and $G(\vec{r}, \vec{r}\,')$ by Eqs. (6.25) - (6.28). The term $4\alpha_s/(3\sqrt{\pi}a)$ is the quark mass renormalization. The one-gluon exchange potential V_x, as in Hartree-Fock, is generally a nonlocal operator. Here, however, since all spatial states are the same, we can write

$$V_x(r) = -\tfrac{2}{3} g_s^2 G(\vec{r}, \vec{r}\,') |\psi_0(\vec{r}\,')|^2 \,. \qquad (8.21)$$

Since $|\psi_0(\vec{r}\,')|^2$ is spherically symmetric, the monopole divergence which appears in E_{self} as $\kappa_v \to 0$ cancels with V_x. These equations are coupled, nonlinear, integro-differential equations. The energy of the nucleon is

$$\mathcal{E} = 3\epsilon - 2g_s^2 \int |\psi_0(\vec{r})|^2 G(\vec{r}, \vec{r}\,') |\psi_0(\vec{r}\,')|^2 f(|\vec{r} - \vec{r}\,'|) d^3r d^3r' \,.$$

We here present the results of a simpler calculation, using the asymptotic form for the self-energy, i.e. we solve

$$\left[\vec{\alpha} \cdot \vec{p} + \beta(m + g_{eff}(\sigma)) - \epsilon_0 \right] \psi_0 = 0 \,, \qquad (8.22)$$

$$-\nabla^2 \sigma + U'(\sigma) + 3g'_{eff}(\sigma) \bar{\psi}_0 \psi_0 = 0 \,, \qquad (8.23)$$

with $m = 0$ and $\kappa(\sigma)$ as given by Eq. (2.11) with $n = 3$. For

$$U(\sigma) = \frac{a}{2} \sigma^2 + \frac{b}{6} \sigma^3 + \frac{c}{24} \sigma^4 + B \,, \qquad (8.24)$$

the parameter values $a = 7.212 \text{ fm}^{-2}$, $b = -805.64 \text{ fm}^{-1}$, $c = 10,000$ and $B = 0.2793 \text{ fm}^{-4} = 55 \text{ MeV/fm}^3$ have been used. This gives $\sigma_v = 0.2222 \text{ fm}^{-1}$ for the vacuum value of the σ-field.

Fig. 8.3 illustrates the properties of the self-consistent bag states with the coupling (8.16). (See also Schuh and Pirner, 1986.) The upper and lower components ($u(r)$ and $v(r)$) of the quark wave function of the lowest energy are plotted together with the sigma field and the effective coupling g_{eff}. Observe the rapid rise of the potential. The corresponding eigenvalue is $\epsilon = 0.36$ GeV, while the total energy of the three-quark system is 1.46 GeV. (This value does not include the recoil corrections.) The glueball mass here is $\sqrt{U''(\sigma_v)} = 1.71$ GeV.

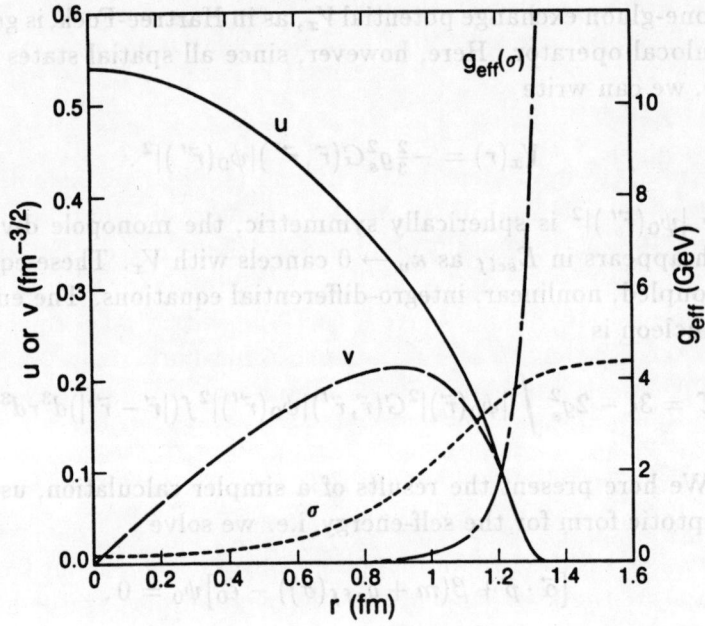

Figure 8.3. Properties of a self-consistent calculation for bag states with the asymptotic coupling. The upper (u) and lower (v) components of the quark wave function and the effective coupling (g_{eff}. right scale) are plotted. The shape of the σ-field is indicated for completeness (arbitrary units). (From Fai, Perry and Wilets, 1988.)

It should be emphasized that the g_{eff} calculated here is to be regarded as a useful approximation for spherical, color-singlet systems. More generally, for nonspherical systems, and to make color-confinement explicit, one must use the self-consistent propagator, as described in Bickeböller, *et al.* (1985) to calculate E_{self} according to Eq. (8.9), or with some other form factor.

The actual calculation of the self-energy was carried out in the one-loop approximation, in the limit of fixed quarks. The computational problem is substantially more difficult for light quarks. However, qualitatively similar results are expected on the basis of a comparison to the results of calculations for light quarks in the MIT bag model (Goldhaber, *et al.*, 1983; Hansson and Jaffe, 1983).

8.3 Massless quarks

Krein, Tang, Wilets and Williams (1988) have calculated the confinement potential in the local, or uniform medium, approximation. Although the local approximation does not exhibit the color confinement inherent to the model (this arises from the bag exterior condition $\kappa \to 0$), it does give a useful approximation to the quark confinement potential and exhibits the chiral invariance.

The quark self-energy in a medium with constant dielectric is

$$\Sigma(p) = \tfrac{16}{3}\pi i \frac{1}{\kappa} \int \bar{d}^4 k\, \alpha_s(k^2) D^{\mu\nu}(k) \gamma_\mu S(p-k) \gamma_\nu \,, \qquad (8.25)$$

where $\alpha_s(k^2)$ is the running coupling constant. $D_{\mu\nu}$ is the free gluon propagator; the κ dependence of the medium propagator has been factored out with the $1/\kappa$. The quark propagator S is defined (for zero mass bare quarks) by

$$S^{-1}(p) = \gamma_\mu p^\mu - \Sigma(p) = A(p^2)\gamma_\nu p^\nu - B(p^2) \,. \qquad (8.26)$$

The notation differs from the reference quoted above by the interchange of A and B.

With zero bare mass quarks, there is no divergence, so no cut-off is required. The scale is set by the QCD scale parameter $\Lambda_{QCD} \approx 200$ MeV/c ≈ 1 fm^{-1}.

In a more recent work, Krein, Tang, Wilets and Williams (1989) have solved the Schwinger-Dyson equation (8.25). Results for the scalar (mass) part of the self-energy as a function of $1/\kappa$ are displayed in Fig. 8.4 for one value of the parameter $\alpha_s(0) = 1.0$. There is a small mass at $\kappa = 1$, a flat region followed by a rapid rise, finally asymptoting to $1/\sqrt{\kappa}$ at very small κ. This is $g(\kappa(\sigma))$; it leads to even more MIT-like behavior than previously assumed forms.

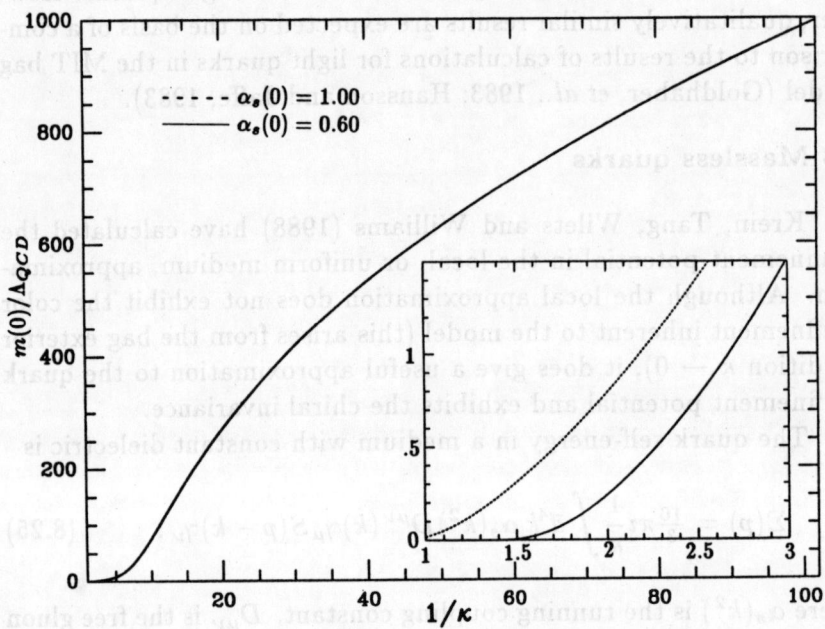

Figure 8.4.   The scalar part of the self-energy (mass) $m(p^2 = 0)$ in units of Λ_{QCD} as a function of the dielectric constant κ. Two values of the input parameter are shown, $\alpha_s(0) = 0.6$ and 1.0. The critical value for dynamical mass generation at $\kappa = 1$ is 0.91. (Krein, Tang, Wilets and Williams, 1989)

Krein, *et al.* (1988), Following Pagels (1976), Cornwall (1980), Delbourgo and Scadron (1979), Scadron (1984), and Govaerts, Mandula and Weyers (1984), use the Ward-Takahashi identity for chirally invariant Lagrangians to obtain a relationship for the quark-axial current vertex function:

$$k^\mu \underset{\longrightarrow}{\Gamma}_{5\mu} (p',p) = \tfrac{1}{2} \underset{\longrightarrow}{\tau} \left[S^{-1}(p')\gamma_5 + \gamma_5 S^{-1}(p) \right], \qquad (8.27)$$

where $k = p' - p$. The underarrow again denotes a vector in isospin space. Using the definition (8.26), one obtains

$$\underset{\longrightarrow}{\Gamma}_{5\mu} (p',p) = \tfrac{1}{2}\gamma_5 \underset{\longrightarrow}{\tau} \left[-2B(p^2)\frac{k_\mu}{k^2} + \cdots \right], \qquad (8.28)$$

where the terms indicated by dots are finite as $k \to 0$. The k^{-2} term dominates as $k \to 0$ and is interpreted as the Nambu-Goldstone boson (pion) (Nambu and Jona-Lasinio, 1961). $B(p^2)$ is the $k = 0$ quark-pion coupling constant; for a hadron bag state it is to be evaluated at $p^2 \approx 0$. Work on the model is continuing. A detailed analysis of the pion structure in the context of an extended Nambu Jona-Lasinio model has been given by Roberts, Cahill and Paschifka (1988).

An alternative approach which uses a confining gluon propagator in place of an explicit chromodielectric function has also recently been considered by Krein, Tang and Williams (1988), Krein and Williams (1989).

Chapter nine

QUANTUM CORRECTIONS AND RENORMALIZATION

9.1 An effective model

The elementary mean field approximation (MFA) considers only valence quarks, while the σ-field is treated classically or in the coherent state approximation. Corrections to the MFA have been termed quantum or loop corrections. Infinities arise in these calculations. If the structure of these terms is the same, up to infinite multiplicative constants, as the original Lagrangian, the infinite constants can be absorbed into renormalized coupling constants or, alternatively, cancelled by similar counter terms. If a finite number of subtractions of counter terms yields a finite result, the model is called renormalizable. This is the case for the Friedberg-Lee model *except* for those terms involving the dielectric function $\kappa(\sigma)$. Such calculations provide an effective Lagrangian for further work.

A serious question arises whether it is appropriate to evaluate quantum corrections in a model which is already an effective theory, and may be presumed to incorporate quantum corrections and renormalization. This is certainly the case for the σ-field, which is not to be regarded as an elementary field, but rather as a condensate of gluons (and quarks). In the models considered here, we consider the quarks to be elementary fields and quantum corrections dress the quarks with virtual sigmas, quark pairs, and gluon excitations. (Note that the gluons already contain some dressing by virtue of the dielectric function.)

From the above point of view, the model Lagrangian need not be renormalizable, and the effective fields (*e.g.* the σ field) should not be further dressed. Rather, parameters of the model Lagrangian are phenomenological, and are to be adjusted at each level of approximation to fit key data. Since there are far more data than parameters, the model has the potential for prediction and/or rejection.

Nevertheless, one can also argue that *changes* in the σ-field loop corrections are relevant. That is, take some reference calculation such as the baryon state. Let us assume that there are counter terms which cancel the loop corrections so that the energy is the same as without the loop corrections. Calculations of other states, such as other hadrons or hadronic collisions should include loop corrections and the counter terms required for the standard calculations.

In spite of the above reservations, we outline here the calculation of the lowest order quantum corrections and renormalizations.

9.2 Uniform system

We consider the Lagrangian defined in Sec 2.1, with the replacement of $\mathcal{L}_\sigma$ by

$$\mathcal{L}_\sigma = \tfrac{1}{2} Z_\sigma \partial_\mu \sigma \, \partial^\mu \sigma - U(\sigma) + U_c(\sigma), \tag{9.1}$$

where

$$U_c(\sigma) = d_c \sigma + \frac{a_c}{2}\sigma^2 + \frac{b_c}{3!}\sigma^3 + \frac{c_c}{4!}\sigma^4 + B_c. \tag{9.2}$$

U_c contains the counter terms which will turn out to be infinite. Z_σ is the wave function renormalization factor.

The quark corrections arise from summing the negative energy states of the Dirac sea in the presence of a classical, uniform scalar potential, say $g_0\sigma_0$.

Let Λ be a momentum cutoff. Then the sea energy is

$$U_f(\sigma_0) = -6 \sum_f \int d^3p\, \theta(\Lambda - |\vec{p}|)\sqrt{p^2 + M_f^2}, \tag{9.3}$$

with $\bar{d}p = dp/2\pi$ and where the spin and color degeneracy are manifested in the factor of 6; the flavors are summed explicitly with $M_f = g_0 + m_f$.

We can then set

$$U_f(\sigma_0) = \frac{3}{\pi^2} \sum_f \int_0^\Lambda p^2 dp \sqrt{p^2 + M_f^2}$$

$$= \frac{3}{8\pi^2} \sum_f \left[\Lambda(2\Lambda^2 + M_f^2)\sqrt{\Lambda^2 + M_f^2} - M_f^4 \ln\left(\Lambda + \sqrt{\Lambda^2 + M_f^2}\right) / |M_f| \right]$$

$$= \lim_{\Lambda \to \infty} \frac{3}{8\pi^2} \sum_f \left[2\Lambda^4 + 2\Lambda^2 M_f^2 + \tfrac{1}{4}M_f^4 - M_f^4 \ln(2\Lambda/|M_f|) \right] \quad (9.4)$$

The terms which are divergent as $\Lambda \to \infty$ are powers up to quartic in σ_0. The coefficients in $U_c(\sigma_0)$ can be chosen to cancel identically the divergent parts of $U_f(\sigma_0)$.

9.3 Nonuniform system

We now consider a nonuniform scalar potential, $g_0\sigma_0(\vec{r})$ which generates a set of eigenvectors ψ_n and eigenvalues ϵ_n (See Fig. 9.1) which can contain both discrete and continuous members. The spectrum is symmetric about zero.

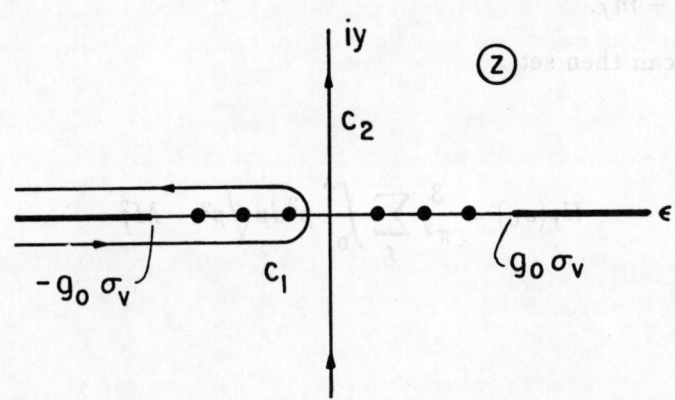

Figure 9.1. The poles of the quark propagator in the complex z-plane. In the Friedberg-Lee model, there are both discrete states and continuum states beginning at $\pm g_0\,\sigma_v$; in the chromodielectric model, there are an infinite number of discrete states and no continuum. The contour C_1 encloses the negative energy poles. Because there are no poles off the real axis, the contour can be deformed to the imaginary axis, C_2.

The sum of the negative energy eigenvalues for each flavor can be expressed as follows:

$$U_f = \sum_{\epsilon_n < 0} \epsilon_n = \frac{1}{2\pi i} \int d^3r \oint_C z\,dz \sum_n \frac{\psi_n(\vec{r})\psi_n^\dagger(\vec{r})}{z - \epsilon_n}$$

$$= \frac{1}{2\pi i}\,\mathrm{Tr}\oint_C z\,dz\,\gamma_0 S(z)\,, \tag{9.5}$$

where

$$(i\vec{\alpha}\cdot\vec{\nabla} - \beta g_0\sigma_0(\vec{r}) + z)\gamma_0 S(\omega; \vec{r}, \vec{r}') = \delta^3(\vec{r} - \vec{r}')\,. \tag{9.6}$$

Here the trace Tr includes the sum over Dirac indices and the spatial integration. It yields the sum over spins. Then

$$S = S_0 + S_0\,g_0\sigma_0 S_0 + S_0\,g_0\sigma_0 S_0\,g_0\sigma_0 S_0 \cdots. \tag{9.7}$$

Terms of the form S_0^n yield numerical coefficients, which are infinite for $n \leq 5$. To bring U_f into a form containing powers of S_0 it is necessary to commute σ_0 with S_0. This introduces terms involving the gradient of $\sigma_0(\vec{r})$, as can be seen by noting

$$[S_0^{-1}, \sigma_0] = i\vec{\gamma} \cdot \vec{\nabla}\sigma_0(\vec{r}). \tag{9.8}$$

This leads to

$$U_f(\vec{r}) = \frac{1}{2\pi i} \oint_C z\, dz \operatorname{Tr} \gamma^0 \{S_0(z) + S_0^2(z) g_0 \sigma_0 + S_0^3(z)(g_0\sigma_0)^2$$

$$+ S_0^4(z)(g_0\sigma_0)^3 + S_0^5(z)[(g_0\sigma_0)^4 - (g_0\nabla\sigma_0)^2] + \cdots\}. \tag{9.9}$$

The first five terms yield a fourth order polynomial in σ with infinite coefficients, exactly as in the uniform case, and can be cancelled by the counter terms, U_c. The coefficient of $(\nabla\sigma_0)^2$ is also infinite and contributes to the wave function renormalization. This plus finite terms allows the replacement $Z_\sigma \to 1$. The remaining terms, represented by the dots, are finite.

In the calculations of Perry (1987), Li (1987), and Li, Wilets and Perry (1988), the Fermi propagator $S(z)$ was calculated. In the expression for the energy, the infinite contributions of (9.9) were explicitly subtracted out before executing the partial wave sums and the contour integral. The contour in Eqs. (9.5) or (9.9) was deformed to run along the imaginary axis: $z = iy$, $y = (-\infty, +\infty)$. This is the prescription of Wichmann and Kroll (1956). Further, highly accurate numerical techniques were developed for evaluating the propagators along the imaginary axis.

Non-uniformity can also be approximated by the gradient expansion, which has been studied in some detail by Li and Perry (1988).

9.4 Bag states: Gaussian moments expansion

In chapter 4 we discussed variation after projection of bag states. We noted that it was essential not to change the model parameters during this procedure. A plane wave representation was employed for the σ-modes. In order to satisfy certain virial theorems, at least approximately, it was necessary to vary also the mode frequencies. This in turn resulted in a variation of the vacuum state from its optimal structure. It was argued that since only differences in the energies between the bag state and the corresponding vacuum state were calculated, the damage done to the vacuum cancelled out.

An alternative approach is to use a representation based on the bag structure (Wilets, 1986). At large distances from the bag, the mode states approach the "true" vacuum states and the vacuum is not damaged. Calculations are tractable at least through one loop corrections in the fermion and quark sectors.

We begin with a localized "bag" state $|B>$ centered at the origin and normalized to $<B|B>= 1$. A state translated by $\vec{Z}$ can be constructed by

$$|\vec{Z}> = e^{-i\vec{P}\cdot\vec{Z}}|B>, \qquad (9.10)$$

where $\vec{P}$ is the total momentum operator and hence the generator of infinitesimal translations. The unnormalized state obtained by projecting $|B>$ onto a state of zero momentum is

$$|\Psi> = V^{-\frac{1}{2}} \int |\vec{Z}> d^3Z. \qquad (9.11)$$

The normalization is given by the integral

$$
\begin{aligned}
<\Psi|\Psi> &= V^{-1} \int d^3Z' \int d^3Z <\vec{Z}'|Z> = \int d^3Z <-\tfrac{1}{2}\vec{Z}|\tfrac{1}{2}\vec{Z}> \\
&= \int d^3Z <B|e^{-i\vec{P}\cdot\vec{Z}}|B> = \int d^3Z <B|j_0(P\,Z)|B> \\
&= \int d^3Z \left[1 - \tfrac{1}{6}<P^2>Z^2 + \tfrac{1}{120}<P^4>Z^4 - \cdots\right], \quad (9.12)
\end{aligned}
$$

where we have used the observation that $|B>$ is a spherically symmetric state and so the angle average can be done immediately. We use the notation $<P^2>=<B|P^2|B>$, etc. Note that $j_0(P\,Z)$ is a function of P^2 and Z^2.

Following the experience of nuclear physics, we re-expand the final square bracket in (9.12) in terms of a Gaussian times a power series:

$$[1 - \tfrac{1}{6} <P^2> Z^2 + \tfrac{1}{120} <P^4> Z^4 - \cdots] =$$

$$e^{-\frac{1}{6}<P^2>Z^2} [1 + (\tfrac{1}{120} <P^4> - \tfrac{1}{72} <P^2>^2)Z^4 + \cdots] . \qquad (9.13)$$

The Gaussian factor was chosen to eliminate the quadratic term in the power series. The $\vec{Z}$-integration can now be performed to yield

$$<\Psi|\Psi> = 2 \left(\frac{6\pi}{<P^2>} \right)^{\frac{3}{2}} \left[1 + \frac{9 <P^4> -15 <P^2>^2}{8 <P^2>^2} + \cdots \right] . \qquad (9.14)$$

A similar procedure can be used to calculate $<\Psi|H|\Psi>$. We note that $\vec{P}$ commutes with the Hamiltonian, so we can write immediately

$$<\Psi|H|\Psi> =$$

$$\int d^3Z \left[<H> - \tfrac{1}{6} <H P^2> Z^2 + \tfrac{1}{120} <H P^4> Z^4 + \cdots\right] . \qquad (9.15)$$

It is convenient and appropriate to extract the same Gaussian from the integrand of (9.15) as was done for the normalization integral. We find

$$<\Psi|H|\Psi> =$$

$$\int d^3Z \, e^{-\frac{1}{6}<P^2>Z^2} \Big[<H> - \tfrac{1}{6}(<H P^2> - <H><P^2>)Z^2$$

$$+(\tfrac{1}{120} <H P^4> - \tfrac{1}{36} <H P^2><P^2>$$

$$+\tfrac{1}{72} <H><P^2>^2)Z^4 + \cdots \Big] . \qquad (9.16)$$

Performing the $\vec{Z}$-integration and taking the ratio to $<\Psi|\Psi>$, we find

$$\frac{<\Psi|H|\Psi>}{<\Psi|\Psi>} = \Bigg[<H> - \frac{3}{2} \frac{<H P^2> - <H><P^2>}{<P^2>}$$

$$+ \frac{9 <H P^4> -30 <H P^2><P^2> +15 <H><P^2>^2}{8 <P^2>^2} + \cdots \Bigg]$$

$$\div \left[1 + \frac{9 <P^4> - 15 <P^2>^2}{8 <P^2 ke2} + \cdots \right]. \tag{9.17}$$

The expression for the energy, or any observable, is thus written in terms of expectation values with respect to localized bag states. In order to define the bag state, we need the Fock space operators associated with $\sigma = \sigma_0(\vec{r})$, π and ψ. As in Chapter 3, we express σ as

$$\sigma(\vec{r}) = \sigma_0(\vec{r}) + \sum_n \sqrt{\frac{1}{2\Omega_n}} \left[a_n + a_n^\dagger \right] u_n(\vec{r}) \equiv \sigma_0 + \sigma_1 \tag{9.18}$$

where $\sigma(\vec{r})$ is a time-independent, c-number, and the conjugate momentum is

$$\pi(\vec{r}) = -i \sum_n \sqrt{\frac{\Omega_n}{2}} \left[a_n - a_n^\dagger \right] u_n(\vec{r}), \tag{9.19}$$

where the $u_n(\vec{r})$ are a complete, orthonormal set and are chosen to be real.

Similarly, the quark field may be expanded as

$$\psi(\vec{r}) = \sum_m c_m \psi_m(\vec{r}), \tag{9.20}$$

where the $\psi_m(\vec{r})$ are a complete set of Dirac spinors. A trial "bag" state $|B>$ is constructed by defining

$$a_n|B> = 0, \qquad \text{for all } n,$$
$$c_m|B> = 0, \qquad \text{for } m \text{ unoccupied,}$$
$$c_m^+|B> = 0, \qquad \text{for } m \text{ occupied.} \tag{9.21}$$

Variation of the expectation value of H, through one loop corrections and with renormalization, leads to the now familiar mean field equations

$$(\vec{\alpha} \cdot \vec{p} + g\beta\sigma_0 - \epsilon_m)\psi_m = 0; \tag{9.22 a}$$

$$-\nabla^2 \sigma_0 + +U'(\sigma_0(\vec{r})) + \sum_{m(valence)} \overline{\psi}_m \psi_m = 0; \qquad (9.22\,b)$$

$$\left[-\nabla^2 + U''(\sigma_0(\vec{r})) - \Omega_n^2\right] u_n = 0; \qquad (9.22\,c)$$

with

$$U(\sigma) = \tfrac{1}{2}a\sigma^2 + \tfrac{1}{3!}b\sigma^3 + \tfrac{1}{4!}\sigma^4$$

and the primes denote differentiation with respect to σ. Evaluation of the one loop sigma energies using contour integrations over Green's functions are described in Wilets (1986).

Note that in Eq. (9.22 c), the Ω_n are eigenvalues for a localized potential $U''(\sigma_0(\vec{r}))$. The spectrum contains both discrete and continuous members. As $n \to \infty$, the functions and frequencies go over to the vacuum values fulfilling the objective of not disturbing the vacuum far from the bag.

Chapter ten

MANY BAG PROBLEM

10.1 Introduction

The description of individual nucleons as three-quark bags or clusters has had considerable success in reproducing elementary properties of hadrons. The treatment of interactions between such structures is more complex. The calculation of N-N scattering has been considered in Sec. 5.3 for soliton bags and by a number of groups for quark potential models. Nuclear matter is more complicated, particularly if one wishes to go beyond the independent pair approximation. Even the Fermi gas is already a very complicated problem for composite systems.

An alternate, albeit simplistic, approach is to consider nuclear matter as a collection of bags like the holes in Swiss cheese, and to study the structure of the system for example, under compression. One expects that the interstices between the bags—the physical vacuum—should disappear as the density is increased leading ultimately to a quark plasma.

Models which place nucleon bags on a regular lattice have received considerable attention. Achtzehnter, Scheid and Wilets (1985) calculated the equation of state for a periodic lattice in the simple soliton model. The sigma field was taken to have the symmetry of a cubic crystal and the quarks were then described by Dirac-Bloch waves. In the mean field approximation without gluons, which was used, the valence quarks occupy one-fourth of the lowest Brillouin band, since each site can accomodate 12 quarks (2 spin, 2 flavor, 3 color) and each site is identified as a baryon consisting of three quarks. Zhang, Derreth, Schäfer and Greiner (1986) solved the prob-

lem using the MIT model on a crystalline lattice. This was a remark-
able feat, since in that model, the quarks are subject to hard bound-
aries with sharp corners. Banerjee, Glendenning and Soni (1985)
employed a hybrid topological soliton model utilizing the Wigner-
Seitz approximation. Reinhardt, Dang and Schulz (1985) also used
the Wigner-Seitz approximation, but with a nontopological soliton
bag model. Goldman and Stephenson (1985) have studied a periodic
quark model in order to understand quark tunneling in nuclei.

Birse, Rehr and Wilets (1988) studied the soliton bag model in
the Wigner-Seitz approximation. Inasmuch as nuclear matter be-
haves more like a fluid than a crystal, they contend that the Wigner-
Seitz approximation is more physical than the crystal model since it
represents an angular average over the location of neighboring sites
and carries no reference to particular crystalline symmetry. This approach has been used very successfully in condensed matter physics
to calculate the equation of state of liquids. We follow here Birse
et al.

A crucial consideration in the calculations is the manner of filling
of the Bloch states within the lowest band. The filling is clear in
the pure MFA without gluons since then the lowest states would be
occupied with unit amplitude up to some Fermi energy determined
by the density. In fact, however, quarks are strongly correlated by
color electric and magnetic forces. Hence levels within a band are
mixed. This mixing leads to an effective probability of occupation
within a band, and has an important effect on the equation of state
and the stability of the system.

The mean field lattice calculations are only a small first beginning from the crystal representation; the subsequent corrections are
probably large and the procedure may not even converge. In the low
density limit, for example, one does not even recover the center-of-
mass corrections discussed at length in Chapter 5. The next step
would be to allow for quantum fluctuations in the σ-field along with
the induced quark excitations.

10.2 The Lattice Model

Consider now the replacement of the moving, fluctuating bags by a regular, periodic face-centered cubic (fcc) lattice of bags. They are characterized by the lattice displacement vectors $\vec{a}_n$. In the mean field approximation, σ is a c-number. Take $\sigma(\vec{r})$ to be periodic in the crystal translation vectors,

$$\sigma(\vec{r}) = \sigma(\vec{r} + \vec{a}_n), \tag{10.1}$$

and to contain the reflectional and discrete rotational symmetries of the fcc lattice.

In the absence of OGE interactions, the quark functions then satisfy Bloch's theorem

$$\psi_{\vec{k}}(\vec{r}) = e^{i\vec{k}\cdot\vec{r}}\phi_{\vec{k}}(\vec{r}), \tag{10.2}$$

where $\vec{k}$ is a continuous vector and $\phi_{\vec{k}}(\vec{r})$ is periodic in the $\vec{a}_n$

$$\phi_{\vec{k}}(\vec{r}) = \phi_{\vec{k}}(\vec{r} + \vec{a}_n), \tag{10.3}$$

although it need not possess the other symmetries of σ. The $\phi_{\vec{k}}$ satisfy the Dirac equation

$$[\vec{\alpha} \cdot (\vec{p} + \vec{k}) + g\,\beta\,\sigma(\vec{r})]\phi_{\vec{k}} = \epsilon_{\vec{k}}\phi_{\vec{k}}, \tag{10.4}$$

where the eigenvalues $\epsilon_{\vec{k}}$ have the characteristic band spectra of the fcc crystal.

At very low density—well separated bags—the self-consistent solutions for σ and ψ are those of isolated bags and the low-lying energy spectrum becomes discrete. As the bags are moved closer together, the eigenvalues $\epsilon_{\vec{k}}$ spread out into bands.

The lattice calculation is feasible, as has been demonstrated by the work of Achtzehnter *et al.* (1985).

10.3 The Wigner-Seitz Cell

Although the Wigner-Seitz cell is a model in its own right, certain boundary conditions associated with it are conveniently determined by reference back to the crystal.

A single "bag" is enclosed in a Wigner-Seitz sphere of radius R such that its volume is the same as that ascribed to each bag in the crystal. (In solid state literature this is usually denoted by r_s.) Because of the assumed spherical symmetry, the lowest band assumes the form for s-states; ψ_k can be represented by

$$\psi_k(\vec{r}) = \begin{pmatrix} u_k(r) \\ i\vec{\sigma} \cdot \hat{r} v_k(r) \end{pmatrix} \chi \qquad (10.5)$$

so that

$$du_k/dr = (g\,\sigma + \epsilon_k)v_k, \qquad (10.6\,a)$$

$$dv_k/dr + 2v_k/r = (-g\,\sigma + \epsilon_k)u_k, \qquad (10.6\,b)$$

and

$$-\nabla^2\sigma(r) + U'(\sigma) + \frac{9g}{\bar{k}^3}\int_0^{\bar{k}} k^2 dk \left[u_k^2(r) - v_k^2(r)\right] = 0, \qquad (10.6\,c)$$

where $\bar{k}$ is the highest k-value in the band that is filled, and is determined below. The factor $9/\bar{k}^3$ assures three quarks per bag irrespective of $\bar{k}$. χ is the spin-flavor-color function. Here the quark functions are normalized to

$$4\pi\int_0^R r^2 dr \left[u_k^2(r) + v_k^2(r)\right] = 1. \qquad (10.7)$$

The boundary conditions on σ are

$$\sigma'(0) = \sigma'(R) = 0. \qquad (10.8)$$

The lowest member of the quark band satisfies the boundary condition

$$u_b'(R) = 0 \quad \Rightarrow \quad v_b(R) = 0. \qquad (10.9)$$

At the top of the band, we have

$$u_t(R) = 0. \qquad (10.10)$$

Using these boundary conditions and a given $\sigma(r)$, one can solve for the corresponding ϵ_b and ϵ_t. The intermediate ϵ's lie in the continuum of the band and do not require the solution of an eigenvalue problem. Rather ϵ_k is specified and (10.6 a, b) are integrated from $r = 0$ to $r = R$ without an eigenvalue search. Let $s = k/k_{top}$ (k_{top} is inversely proportional to the lattice spacing); $s \leq 1$. They make the ansatz that

$$\epsilon(s) = \epsilon_b + (\epsilon_s - \epsilon_b) \sin^2(\pi s/2). \qquad (10.11)$$

Using the reduced momentum label s instead of k, the inhomogeneous term in (10.6 c) can now be written

$$\frac{9g}{\bar{s}^3} \int_0^{\bar{s}} s^2 ds \left[u_s^2 - v_s^2 \right]. \qquad (10.12)$$

10.4 Band filling

Each level is 2 (spin) $\times$ 2 (flavor) $\times$ 3 (color) = 12-fold degenerate. If one were to neglect gluonic interactions, the lowest band would be densely-filled one-fourth of the way up, implying $\bar{s} = (\frac{1}{4})^{1/3}$. However, the nucleon and delta are each particular linear combinations of products of single quark states, and one is interested in the nucleon states.

For isolated bags, there are

$$\binom{12}{3} = 220$$

three-quark states which can be constructed from spacially identical orbitals. Of these, there are only 20 color-singlet states: the 4 nucleon and 16 delta states. The soliton model guarantees color-confinement for isolated bags, although color-percolation can occur for overlapping bags. The color-electric matrix elements (required for color-confinement) are much greater than $\alpha_s / <r>^{1/2}$, which is several hundred MeV. The color-magnetic interaction is responsible for at least half of the N-Δ splitting of 293 MeV; the rest can be interpreted as pionic interaction (which must also be included in the following argument). In the case of band structure in the crystal or Wigner-Seitz models, these gluonic matrix elements lead to a mixing of the band members and a separation of the many-particle states into nucleon bands, delta bands, and bands of different color symmetry. They find that the width of the lowest band varies from zero (for well-separated bags) to about 400 MeV at $R = 0.8$ fm, which corresponds to about 3.4 times normal nuclear density. For large separations the mixing matrix elements are large compared to the band width; for small separations they may become comparable. They are thus led to the assumption that band levels are fully mixed by the color-electric and magnetic forces, and that all levels within a band have equal probability of occupation. This corresponds to setting $\bar{s} = 1$.

Note that the next band, corresponding to the first excited s level, lies, in the MIT model, at 2.645 times the lowest s-level, roughly 500 MeV above the lowest level. Hence they ignore mixing to the higher bands.

A similar approach has been used by Kerman and Dagdeviren (1986) to produce 3-quark correlations in the plasma phase. There the calculation of gluon-exchange energies is simplified by the fact that the Block waves are simply plane waves.

10.5 The Uniform Plasma

In the high density limit, a uniform plasma is the preferred phase. In the MFA, the energy per unit volume of this uniform plasma is given by $(\bar{d}^3 k = dk^3/(2\pi)^3)$

$$\frac{E}{V} = 12 \int \bar{d}^3 k \left[k^2 + (g\sigma) \right]^{1/2} \theta(k_F - k) + U(\sigma)$$

$$= \frac{3}{4\pi^2} \left[2k_F \epsilon_F^3 - (g\sigma)^2 k_F \epsilon_F - (g\sigma)^2 \ln\left(\frac{k_F + \epsilon_F}{g\sigma} \right) \right] + U(\sigma),$$
$$(10.13)$$

with $\epsilon_F = \left[k_F^2 + (g\sigma)^2 \right]^{1/2}$; 12 is the degeneracy factor. The baryon density, which is one-third of the quark density, is given by

$$\frac{N}{V} = \left(\tfrac{4}{3}\pi R^3\right)^{-1} = 4 \int d^3k\, \theta(k_F - k) = \frac{2}{3\pi^2}k_F^3, \qquad (10.14)$$

or

$$k_F = \tfrac{1}{2}(9\pi)^{1/3} R^{-1}. \qquad (10.15)$$

The energy (10.13) must be minimized with respect to σ for fixed volume. Since $U(\sigma)$ has a local minimum at $\sigma = 0$ and the integral has an absolute minimum there, it follows that $\sigma = 0$ is always a local minimum (and usually the lowest minimum) of the full energy density. Therefore

$$\frac{E}{V} = \frac{3}{2\pi^2}k_F^4 + B \qquad (10.16)$$

or

$$\frac{E}{N} = \tfrac{9}{4}k_F + \frac{3\pi^2}{2k_F^3}B = 3.4273R^{-1} + \tfrac{4}{3}\pi R^3 B. \qquad (10.17)$$

Note that (10.17) depends only on the bag constant B and the cell radius R. It is of the same *form* as the energy expression in the MIT model, except that here the coefficient of R^{-1} is 3.4273 compared with the MIT value (for three quarks) of $3 \times 2.0428 = 6.1284$. Thus in the MIT model the uniform plasma has a lower energy per baryon than isolated bags when gluonic effects are ignored. This is still true if the Casimir term (see Sec. 1.4) is included.

Gluonic effects are needed if the model is to provide more than a crude qualitative description of the hadronic phase. Similarly one must include gluonic contributions to the energy of the plasma. These have been calculated by Freedman and McLerran (1977) and by Baluni (1978). To order $\alpha_s \ln \alpha_s$, the plasma energy is

$$\frac{E}{V} = \frac{3}{2\pi^2}k_F^4 \left[1 + \frac{2\alpha_s}{3\pi} + \frac{\alpha_s^3}{3\pi^3}\left(2\ln(2\alpha_s/\pi) + 6.79\right)\right] + B, \qquad (10.18)$$

where α_s is the strong coupling constant appropriate to the Fermi momentum. The leading-logarithm expression for α_s is

$$\alpha_s(k_F) = \frac{12\pi}{29 \ln(k_F^2/\Lambda^2)}; \qquad (10.19)$$

$\Lambda = 150$ MeV was used for the QCD scale parameter.

10.6 Numerical Results

Calculations based on several sets of parameters were performed; all belonged to the family $a = 0$. The results did not differ appreciably over a wide range of values of c, except that solutions are "lost" sooner (i.e. lower density) for the smaller values of c. For $c = 10^5$, for example, solutions were obtained for all values of R; for $c = 10^4$, the solution was lost somewhere in the range $0.8 < R < 0.9$ fm. For calculations reported here, the parameter set is

$a = 0.0$

$b = -700.43$ fm^{-1}

$c = 10^4$

$g = 10.98$

from which it follows that $B = 0.27$ fm$^{-4} = 53$ MeV/fm^3, $g\sigma_v = 2.3$ fm$^{-1} = 454$ MeV, $m_{GB} = 8.58$ fm$^{-1} = 171$ MeV.

Stable Wigner-Seitz solutions could not be obtained, even at the radius of normal nuclear density ($R_0 = 1.12$ fm), with fillings corresponding to $\bar{s}$ significantly less than unity. In such cases, the energy was found to decrease with decreasing R. Since these calculations are based on the MFA, they do not contain the higher order terms necessary to describe recoil corrections nor details of the N-N interaction, namely the effective attraction attributed to one and two pion (or "sigma meson") exchange. The energies are too high at low density inasmuch as the parameters had been fitted to yield the mean nucleon-delta mass including recoil corrections.

The quark density (normalized to unity per cell) and the σ functions are displayed in Fig. 10.1. In these calculations, the functions shrink as R is reduced, but the value of σ at the cell boundary does not decrease as we might have expected. This means that the dielectric function essentially vanishes across the boundary and gluons are confined to their own cell. Individual quarks cannot percolate, but color-singlet clusters of three can.

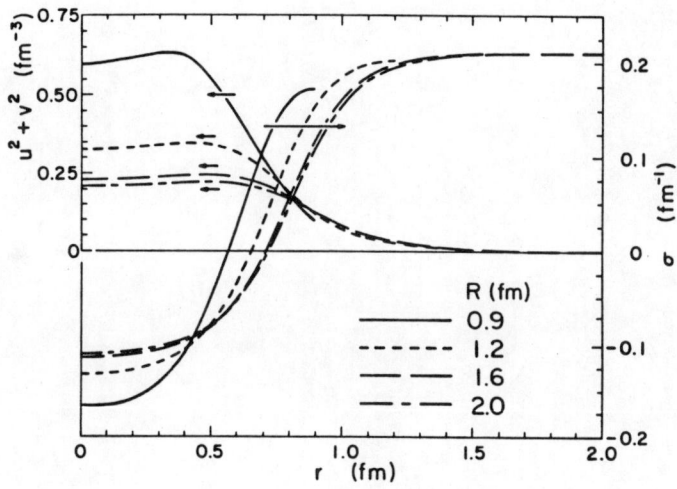

Figure 10.1. The quark densities (normalized to one per cell, left-hand scale), and the sigma function (right-hand scale) are plotted against r for several cell sizes.

On the basis of the simple quark plasma model and empirical data, they drew the following conclusions which are, however, dependent on the bag constant and the QCD scale parameter: There is a first order phase transition at zero temperature. The two phases in equilibrium correspond to a nuclear matter phase at roughly six times normal density and a plasma phase of roughly twelve times normal nuclear density. Heavy ion collisions have already achieved densities of about four times normal nuclear matter without a signal of a phase transition.

It would be interesting to redo the calculation using the chromodielectric model described in Chapter 8, calculating the electric self energy consistently with $\kappa(\sigma)$.

Chapter eleven

RETROSPECT AND PROSPECTS

The general soliton model Lagrangian described in Chapter 2 is the basic QCD Lagrangian supplemented by a dynamic scalar (σ) field which moderates the dielectric properties of the medium. The sigma field is interpreted as a gluon and quark pair condensate. Model parameters in $U(\sigma)$ and the dielectric function $\kappa(\sigma)$ are to be adjusted at each level of sophistication of the calculations in order to reproduce key physical data and to prevent double counting, the latter because exact calculations should already account for the collective modes the sigma field simulates. Ultimately, one should have a decoupling or disappearance of the sigma field. We are still far from that stage in QCD calculations, so we must regard the model as phenomenological, and look to experiment and fundamental theory for guidance in determining parameters, structure of the model, and indeed for verification or rejection of the model. The only fundamental QCD calculations in the confinement regime come from lattice gauge theory. Although to date these have not been of sufficient precision or reliability to give decisive guidance, we look forward to continued development of the field.

The Friedberg-Lee nontopological soliton model has been used effectively to describe the low energy structure and dynamics, including collisions, of hadrons. The model yields absolute confinement in color singlet states by means of the self and mutual interaction of quarks through the exchange of gluons. Most of the calculations described here do not utilize the absolute confinement mechanism. For such calculations, quark excitations are necessarily limited to a few hundred MeV before one encounters the quark continuum. Several versions of "color-dielectric" models assume absolute spatial confinement and have been used with some success for calculations where the color singlet nature of the wave function has been imposed "by hand."

An important feature of the soliton model is the ability to handle dynamics. This is needed for the calculation of N-N and $\overline{N}$-N collisions, nucleon-meson, etc.

Consider the N-N interaction problem, which is central to nuclear physics. All of the needed pieces for the complete calculation have been implemented by various programs already discussed: The role of generator coordinates (including internal dynamics), one gluon exchange (using gluon propagators in the dielectric medium) and the tacking on of the long-range one-boson exchange potential. With more complexity, one could study three-nucleon systems and extract three-body forces. The results could be incorporated into conventional many-body calculations. As also discussed, the many-nucleon problem can be attacked beginning with a crystal structure; the next step wouled be to include oscillations of the component bags.

In pushing the domain of applicability to higher energies, one needs to look more carefully at other features of QCD. For this, the chromodielectric model holds much promise. By simply removing the direct quark-sigma coupling term, one obtains a model Lagrangian which is chirally invariant for zero current mass quarks. Confinement is still effected by the chromoelectric self energy which, however, must be calculated explicitly. For this the running of the coupling constant and asymptotic freedom play a crucial role. The pion emerges as Nambu-Goldstone boson. The resulting picture takes on some features of the cloudy bag model, but contains dynamics.

The powerful methods which have been applied to the Friedberg-Lee model can now be applied to the chromodielectric model to calculate hadronic spectra and dynamics at higher energy, and the nuclear many-body problem as a function of temperature with the transition to a quark-gluon plasma. Indeed, the goal is a general description of nuclear and particle physics at low and intermediate energies, *i.e.* up to many GeV.

Appendix

NUMERICAL METHODS

A.1 Introduction

Numerous physical problems fall into the category of coupled nonlinear differential equations subject to boundary conditions equal in number to the number of first order equations. We will consider here the Friedberg-Lee nontopological model in the mean field approximation. Methods are discussed for solving coupled non-linear differential equations, the Dirac eigenvalue problem and nonlinear integro-differential equations.

Recall the Hamiltonian (3.1). In the MFA, we treat $\sigma(\vec{r})$ as a c-number and consider only valence quarks. Extremization of the expectation value of H with respect to valence quark functions $\psi_k^\dagger$ (subject to the normalization constraint) and with respect to σ leads to the set of coupled equations (3.8), which for the spherical case reduce to Eqs. (3.12):

$$\left(\frac{d}{dr} + \frac{(\kappa+1)}{r}\right) u_n + (g(\sigma) + \epsilon_n)\, v_n = 0, \qquad (A.1a)$$

$$\left(\frac{d}{dr} - \frac{(\kappa-1)}{r}\right) v_n + (g(\sigma) - \epsilon_n)\, u_n = 0, \qquad (A.1b)$$

$$-\frac{1}{r}\frac{d^2}{dr^2}\, r\sigma + U'(\sigma) + \frac{g'(\sigma)}{4\pi} \sum_{n(valence)} \left(u_n^2 - v_n^2\right) = 0\,, \qquad (A.1c)$$

with the normalization condition

$$\int_0^\infty r^2 dr\, \left(u_n^2 + v_n^2\right) = 1. \qquad (A.1d)$$

We have assumed here that the quark distribution is also spherically symmetric. The generalization to non-spherical systems has also been effected (Schuh, 1985).

We now consider various techniques for solving Eqs. (A.1) or their non-spherical generalizations.

A.2 Non-linear differential equations

Eqs. (A.1 a, b) represent a linear eigenvalue problem, for which numerous methods are available (*c.f.* Goldflam and Wilets, 1982; Horn, *et al.*, 1986) Eq. (A.1 c) is a second order nonlinear differential equation coupled to the solutions of (A.1 a, b, d).

The standard methods for solving nonlinear differential equations are variations of the well-known Newton-Raphson method for nonlinear algebraic equations. This consists of linearization about an approximate solution:

Consider first the algebraic problem

$$F_i(y_1 \cdots y_I) \equiv F_i(y_j) = 0, \qquad i = 1 \cdots I. \qquad (A.2)$$

Let $\{y_j^n\}$ be the n-th approximation to the set $\{y_j\}$, and set

$$y_j^{n+1} = y_j^n + \delta_j^n . \qquad (A.3)$$

Expansion of (A.2) in a Taylor series gives

$$F_i(y_j^n) + \sum_j \delta_j^n \frac{\partial}{\partial y_j} F_i(y_j^n) + \cdots = 0 . \qquad (A.4)$$

Eq. (A.4) can be solved through first order in δ_j^n by a linear equation solver, and the process iterated until convergence.

Henyey and Wilets (Berg and Wilets, 1956; Henyey, Wilets, Böhm, Le Levier and Levee, 1959) utilized the Newton-Raphson principle to develop a method for solving coupled nonlinear differential equations *subject to boundary conditions,* and applied it first to the problem of stellar structure. Let the system of equations be

$$y_i'(x) + f_i(y_j(x)) = 0 . \qquad (A.5)$$

In analogy with the algebraic case, let $\{y_i^n(x)\}$ be the n-th approximation to the set $\{y_i(x)\}$, and set

$$y_i^{n+1}(x) = y_i^n(x) + \delta_i^n(x). \qquad (A.6)$$

Expanding the dependent variables about each point x, we have

$$\delta_i^{n\prime}(x) + \sum_j \delta_j^n \frac{\partial}{\partial y_j} f_i(y_j^n(x)) = -\left[y_i^{n\prime}(x) + f_i(y_j^n(x))\right] + \cdots. \quad (A.7)$$

Ignoring higher order terms, (A.7) is a set of coupled, first order, linear, inhomogeneous differential equations, subject to appropriate boundary conditions. They can be solved for the $\delta_i^n(x)$, the y_i^{n+1} obtained, and the process iterated until convergence. Convergence is quadratic—the number of significant figures doubles with each iteration.

The essential feature of the method is that the boundary conditions are satisfied at every iteration. The equations are very stable so long as one is in the neighborhood of a solution. The principle here is what I call "the girl with a curl" syndrome:

> There was a little girl,
> And she had a little curl
>> Right in the middle of her forehead.
> When she was good,
> She was very, very good,
>> And when she was bad she was horrid.

> *H. W. Longfellow*

The set (A.6) can be solved by standard linear, inhomogeneous differential equation solver routines. One can (for example) obtain one particular solution of the inhomogeneous differential equation which satisfies the inside boundary conditions and I solutions of the homogeneous differential equations which also satisfy the inside boundary conditions. The most general solution is a linear combination of the particular solution and an arbitrary linear combination of the homogeneous solutions. This gives I constants. The same can be done satisfying the outside boundary conditions. Now we have $2I$ constants. Fitting the inside and outside functions and their derivatives at an intermediate match point fixes the $2I$ constants. The programs COLSYS (Ascher, Christiansen and Russel, 1979) and LARRY (Carson, 1984) utilize this principle but differ in the numerical techniques.

An alternative method is to express the differential equations as difference equations on a fixed grid. Solution of the inhomogeneous difference equations involves the solution (inversion) of a band matrix problem, for which well-known techniques are available (cf. Goldflam and Wilets, 1982, and Horn *et al*, 1986).

Different procedures have been used for solving the full set of equations (A.1). Goldflam and Wilets solve the eigenvalue equations (a, b), assert the normalization condition (d), and then solve the nonlinear sigma equation (c). The overall process is iterated, with a convergence (or damping) factor of 0.5 on the sigma increment, to the point of convergence.

Alternatively, Saly (1983), Saly and Sundaresan (1984), Köppel and Harvey (1985) have employed variations of the following: Consider (A.1 a, b, c) to be a set of three nonlinear equations to be solved simultaneously. For fixed g_0 and ϵ, a solution can be obtained which, however, does not satisfy the normalization condition (d). Then a search can be made on ϵ until (d) is satisfied. Or, by scaling, one can interpret any solution of the set as a normalized solution with *some* coupling constant g_0. This is particularly useful if one is scanning a parameter set.

Dodd and Lohe (1985) cleverly are able to fully specify the model parameters and solve the full set of equations simultaneously. They do this by augmenting the variables with a normalization function, say

$$\xi(r) = \int_0^r [(u^2(r') + v^2(r'))]r'^2 dr' . \qquad (A.7)$$

Then

$$\xi' = r^2(u^2 + v^2) , \qquad (A.8)$$

with $\xi(0) = 0$ and $\xi(\infty) = 1$. The eigenvalue is also introduced as a variable, $\epsilon = \epsilon(r)$, satisfying the deceptively trivial differential equation

$$\epsilon'(r) = 0 \qquad (A.9)$$

which is nevertheless a member of the coupled set. Then these 6 coupled, nonlinear, first order equations are solved simultaneously. From my own experience, I would judge that this probably requires the order of ten to twenty iterations, depending on the quality of the initial guess.

A.3 Nonspherical mean field solutions

The first obvious extension of these system of equations is to nonspherical systems, such as are encountered in hadron-hadron scattering. One way of handling such systems is by expansion in spherical harmonics. Consider the case of axial and reflectional symmetry. The σ-field can be expanded in a set of even Legendre polynomials,

$$\sigma(\vec{r}) = \sum_{l=0}^{l_m} \sigma_l(r) P_l(\cos\theta). \qquad (A.10)$$

This multiplies the number of the sigma equations (A.1 c) by $\frac{1}{2}l_m + 1$, where l_m is the maximum order of l. The Dirac function becomes a linear combination of κ-eigenfunctions:

$$\psi_n = \sum_\kappa \begin{pmatrix} u_n^\kappa(r) \\ i\vec{\sigma}\cdot\hat{r}\, v_n^\kappa(r) \end{pmatrix} \mathcal{Y}_{\kappa m}. \qquad (A.11)$$

Again going up to l_m in orbital states, there are $2l_m + 1$ states for $|m| = \frac{1}{2}$, decreasing with $|m|$, approximately half of which are even parity and half are odd parity. Although the different parity states do not *mix*, they do *couple*. For n spatially distinct quark states, the number of coupled, first order differential equations to be solved is the order of $l_m + 2l_m n + 2n$. Although n is usually small, like 2 or 4, l_m is characteristically 10, so we could be dealing with about 50 equations.

The mean field solutions provide a basis for generator coordinate calculations. By employing a coherent state representation of the sigma field, one can derive a Schrödinger-type equation for the collective (*i.e.* interparticle) coordinate.

A.4 Gluon propagator

In the Abelian approximation, the gluon propagator is identical to the Maxwell Green's function except for the appearance of the $SU(3)$ color matrices. In the soliton model, however, one must solve for the Green's function in a dielectric medium which contains a spatially dependent $\kappa(\vec{r})$. This again involves the solution of coupled differential equations. The one gluon exchange magnetic interaction can be treated in perturbation theory, so I will not consider it further here. (See Bickeböller, 1984).

Of special interest is the electric interaction, since this is the mechanism of color confinement. In transverse gauge $\vec{D} = \kappa \vec{E}$ satisfies Gauss's law,

$$\vec{\nabla} \cdot \vec{D} = \rho. \qquad (A.12)$$

The scalar Green's function $G_{00} = G^{00} \equiv G$ satisfies the time-independent differential equation

$$-\vec{\nabla} \cdot \kappa(\vec{r})\vec{\nabla}G(\vec{r},\vec{r}') = \delta^3(\vec{r} - \vec{r}'). \qquad (A.13)$$

To simplify this equation one can define

$$\overline{G}(\vec{r},\vec{r}') \equiv \sqrt{\kappa(\vec{r})}\, G(\vec{r},\vec{r}')\sqrt{\kappa(\vec{r}')}, \qquad (A.14)$$

where $\overline{G}$ satisfies

$$(-\nabla^2 + W(\vec{r}))\overline{G}(\vec{r},\vec{r}') = \delta^3(\vec{r} - \vec{r}'), \qquad (A.15)$$

with the "potential" $W(\vec{r})$ given by

$$W(\vec{r}) \equiv \tfrac{1}{4}|\vec{\nabla}\ln\kappa(\vec{r})|^2 + \tfrac{1}{2}\nabla^2\ln\kappa(\vec{r}). \qquad (A.16)$$

We expand $W(\vec{r})$ and $\delta^3(\vec{r} - \vec{r}')$ into spherical harmonics,

$$W(\vec{r}) = W_{LM}(r)Y_{LM}(\Omega), \qquad (A.17)$$

$$\delta^3(\vec{r} - \vec{r}') = \frac{1}{r^2}\delta(r - r')Y_{LM}(\Omega)Y_{LM}^*(\Omega'), \qquad (A.18)$$

where the repeated index summation convention is used.

For the scalar Green function we make the ansatz

$$\overline{G}(\vec{r},\vec{r}') = C_{\alpha\alpha'} \frac{1}{rr'} f^{\alpha}_{lm}(r_<) Y_{lm}(\Omega_<) g^{\alpha'}_{l'm'}(r_>) Y^*_{l'm'}(\Omega_>), \quad (A.19)$$

where the quantities $(r_<, \Omega_<, i_<)$ refer to (r, Ω, i) if $r > r'$ and to (r', Ω', i') if $r < r'$; $(r_>, \Omega_>, i_>)$ is defined correspondingly. The radial functions $f^{\alpha'}_{lm}(r)$ and $g^{\alpha}_{lm}(r)$ satisfy the coupled differential equations

$$\left\{\left(-\frac{d^2}{dr^2} + \frac{l(l+1)}{r^2}\right)\delta_{ll'}\delta_{mm'} + W_{LM}(r) < lm \mid Y_{LM} \mid l'm' >\right\}$$

$$\times \left\{\begin{array}{c} f^{\alpha}_{l'm'}(r) \\ g^{\alpha'}_{l'm'}(r) \end{array}\right\} = 0. \quad (A.20)$$

The set $\{\alpha\}$ of solutions, which are regular at the origin, is given by $\{f^{\alpha}_{lm}\}$ and the set $\{\alpha'\}$, which are regular at infinity, by $\{g^{\alpha'}_{l'm'}\}$. If $\kappa(\vec{r})$ goes asymptotically to a *constant greater than zero*, the corresponding boundary conditions are

$$f^{\alpha}_{lm}(r) \sim r^{l+1}\delta^{\alpha}_{lm} \quad \text{for } r \to 0, \quad (A.21\,a)$$

$$g^{\alpha'}_{l'm'}(r) \sim \frac{1}{r^{l'}} \delta^{\alpha'}_{l'm'} \quad \text{for } r \to \infty. \quad (A.21\,b)$$

If there are no symmetries in the problem, all (l, m) are coupled. In an axially symmetric problem, the potential W can be expanded in terms with the magnetic quantum number $M = 0$ only, and the differential equations (A.20) decouple for different m. For reflectional symmetry, only even L appear. The coefficients $C_{\alpha\alpha'}$ in (A.19) satisfy the linear algebraic equations

$$\left\{f^{\alpha}_{lm}(r)n^{\alpha'}_{l'm'}(r) - j^{\alpha}_{l'm'}(r)n^{\alpha'}_{lm}(r)\right\}C_{\alpha\alpha'} = 0, \quad (A.22\,a)$$

$$\left\{\left(\frac{d}{dr}f^{\alpha}_{lm}(r)\right)n^{\alpha'}_{l'm'}(r) - j^{\alpha}_{l'm'}(r)\left(\frac{d}{dr}n^{\alpha'}_{lm}(r)\right)\right\}C_{\alpha\alpha'} = \delta_{l,l'}\delta_{m,m'},$$

$$(A.22\,b)$$

at any radius r. Equations (A.22 b) with $l < l'$ are redundant, so that the set (A.22) reduces to the linearly independent set of equations: $l < l'$ for (A.22 a) and $l \geq l'$ for (A.22 b). After solving the linear equations for $C_{\alpha\alpha'}$, the scalar Green function is completely determined.

REFERENCES

Abramowitz, M. and Stegun, I. A. (1965) *Handbook of Mathematical Functions*, (Dover, New York).

Achtzehnter, J., Scheid, W. and Wilets, L. (1985) Phys. Rev. **D 32**, 2414.

Achtzehnter, J. (1988) "Is Nucleon a Dirac Particle?" Doctoral Dissertation, Univ. of Washington.

Achtzehnter, J. and Wilets, L. (1988) Phys. Rev. **C 38**, 5.

Anastasio, M. R., Celenza, L. S., Pong, W. S. and Shakin, C. M. (1983) Phys. Reports, **100**, 327.

Ascher, U., Christiansen, J. and Russell, R. D. (1979) Math. Comp. **33**, 659; *Lecture Notes in Computer Science*, **76**, (Springer, N. Y.);

—— (1981) ACM. Trans. Math. Software **7**, 223.

—— Hake, J.-Fr. (1986) ACM. Trans. Math. Software **12**, 283.

Baker, M., Ball, J. S. and Zachariasen, F. (1986) Phys. Rev. **D 34**, 3894;

—— (1988) Phys. Rev. **D 37**, 1036; Phys. Rev. Lett. **61**, 521.

Baluni, V. (1978) Phys. Rev. **D 17**, 2092.

Banerjee, B., Glendenning, N. K. and Soni, V. (1985) Phys. Lett. **155 B**, 213;

—— Glendenning, N. K. and Banerjee, B. (1986) Phys. Rev. **C 34**, 1072

Bardeen, W. A., Drell, S. D., Weinstein, M. and Yan, T.-M. (1975) Phys. Rev. **D 11**, 1094.

Bayer, L., Forkel, H. and Weise, W. (1986) Z. Physik **A324**, 365.

Berg, R. and Wilets, L. (1955) Proc. Phys. Soc. London **A68**, 229.

Betz, M. and Goldflam, R. (1983) Phys. Rev. **D28**, 2848.

Bickeböller, M., (1984) 'Der Gluon-Propagator im Soliton-Bag-Modell,' Diplom Thesis, Univ. of Bonn;

Bickeböller, M. (1986) "The Gluon Propagator in the Soliton Bag Model and its Applications," Doctoral Dissertation, University of Washington;

—— Bickeböller, M., Birse, M. C. and Wilets, L. (1987) Zeit. für Physik **A 326**, 89.

—— Bickeböller, M., Goldflam, R. and Wilets, L. (1985) J. Math. Phys. **26**, 1810.

Bickeböller, M., Birse, M. C. Marschall, H., and Wilets, L. (1985) Phys. Rev. **D 31**, 2892.

Birse, M., Rehr J. J., and Wilets L. (1988) Phys. Rev. **C 38**, 359.

Bjorken, J. D. and Drell, S. D. (1964) *Relativisitic Quantum Mechanics* (McGraw-Hill, New York).

Bleszynski, E., Bleszynski, M. and Jaroszewicz, T. (1987) Phys. Rev. Lett. **59** 423.

Bolsterli, M. (1979) Adv. Nucl. Phys. **11**, 367 (1979);

—— (1983) Phys. Rev. **D27**, 349.

Brodsky, S. J. (1984) Comm. Nucl. Part. Phys., **12**, 213.

Broniowski, W., Cohen, T. D. and Banerjee, M. K. (1987) Phys. Lett. **187 B**, 229.

Brown, G. E., Rho, M. and Vento, V. (1979) Phys. Lett. **84 B**, 383.

Brown, W. D., Puff, R. D. and Wilets, L. (1970) Phys. Rev. C **2**, 331.

Buchmüller, W, (1982) Phys. Lett. **112 B**, 479.

Cahill, R. T. and Roberts, C. D. (1985) Phys. Rev. D. **32**, 2419.

Cahill, R. T., Roberts, C. D. and Praschifka, J. (1987) Phys. Rev. D. **36**, 2804.

—— Cahill, R. T., "Hadronisation of QCD", Flinders Univ. of South Australia preprint FPPG-R-13-88.

Carson, L. C. (1986), private communication.

Casimir, H. B. G. (1948) Proc. Kon. Ned. Akad. Wetenschap. B **51**, 793.

Casimir, H. B. G. and Polder, D. (1948) Phys. Rev. **73**, 360.

Chanfray, G., Nachtmann, O. and Pirner, H. J. (1984) Phys. Lett. **147B** , 249;

—— Nachtmann, O. and Pirner, H. J., (1984)Z. Physik **C 21**, 277.

Chodos, A, Jaffe, R. L., Johnson, K., Thorn, C. B. and Weisskopf, V., F. (1974) Phys. Rev. **D9** 3471;

—— Chodos, A., Jaffe, R. L., Johnson, K. and Thorn, C. B. (1974) Phys. Rev. **D10**, 2599.

Chodos A. and Thorn, C. B. (1975) Phys. Rev. **D12**, 2733.

Clark, B. C., Hama, S., Mercer, L. R., Ray, L. and Serot, B. D. (1983), Phys. Rev. Lett., **50**, 1644.

Cornwall, J. M. (1980) Phys. Rev. D **22**, 1452.

Crawford, G., Dethier, J-L., Wilets, L. and Alberg, M. (1985) Proc. Int. Conf. on Antinucleon- and Nucleon-Nucleus Interactions, Telluride, Colorado.

Crawford, G. (1986) "Quarks and the Saturation Properties of Nuclear Matter", Doctoral Thesis, Univ. of Washington.

Crawford, G. and Miller, G. A. (1987) Phys. Rev. C **36**, 1956.

da Providencia, J. (1973) Nucl. Phys. **B 57**, 536.

da Providencia, J. and Urbano, J. (1978) Phys. Rev. **D 18**, 4208.

DeGrand, T., Jaffe, R. L., Johnson, K. and Kiskis, J. (1975) Phys. Rev. **D 12**, 2060.

Delbourgo R, and Scadron, M.D. (1979) J. Phys. G **5**, 1621.

DeTar, C. (1978) Phys. Rev. **D 17**, 302 and 323.

Dethier, J.-L., Goldflam, R., Henley, E. M.and Wilets, L. (1983) Phys. Rev. **D27**, 2193.

Dethier, J.-L. (1985) "A Soliton Bag Model of the Nucleon and the Delta Dressed by a Quark-Antiquark Pion," Doctoral Dissertation, Univ. of Washington;

—— Dethier, J.-L. and Wilets, L. (1985) Phys. Rev. **D 34**, 207.

Dodd, L. R. and Lohe, M. A. (1985) Phys. Rev. **D 32**, 1816.

Dodd, L. R. and Williams, A. G. (1988 a) Phys. Rev. D **37**, 1971.

Dodd, L. R. and Williams, A. G. (1988 b) Phys. Lett. **210 B**, 10.

Dodd, L. R., Williams, A. G. and Thomas, A. W. (1987) Phys. Rev. D **35**, 1040.

Donoghue, J. F. and Johnson, K. (1980) Phys. Rev. **D21**, 1975.

Eichten, E., Gottfried, K., Kinoshita, T., Lane, K. D. and Yan, T. M. (1980) Phys. Rev. **D 21**, 203.

Fai, G., Perry. R. and Wilets, L. (1988), Phys. Lett. **B 208**, 1; erratum, *ibid* **B**.

Feinberg, G. and Sucher, J. (1979) Phys. Rev. D **20**, 1717.

Fiebig, H. R. and Hadjimichael, E. (1984) Phys. Rev. D **30**, 181.

Fokker, A. D. (1929) *Relativiteitstheorie* (Groningen: P. Noordhoff).

Foldy, L. L. (1952) Phys. Rev **87**, 688.

Foldy, L. L. and Wouthuysen, S. A. (1949) Phys. Rev. **78**, 29.

Freedman, B. A. and McLerran, L. D. (1977) Phys. Rev. **D 16**, 1130 and 1169.

Friedberg, R., Lee, T. D. and Sirlin, A. (1976) Phys. Rev. **D 13** 2739;
—— Nucl. Phys. **B115**, 1, 32.

Friedberg, R. and Lee, T. D. (1977) Phys. Rev. **D15**, 1694; **D16**, 1096.

Friedberg, R. and Lee, T. D. (1978) **D18**, 2623.

Fujiwara, I. (1959) Prog. Theor. Phys. **21**, 902.

Gasser, J. and Leutwyler, H. (1982) Phys. Rep. **87 C**, 78.

Gell-Mann, M. (1964) Phys. Lett. **8**, 214.

Glauber, R. J. (1963) Phys. Rev. **131**, 2766.

Glendenning, N. K. (1986) Phys. Rev. Lett. **57**, 1120.

Goldflam, R. and Wilets, L. (1982) Phys. Rev. **D25**, 1951.

Goldhaber, S. N., Hansson, T. H., and Jaffe, R. L. (1983) Phys. Lett. **133 B**, 445.

Goldman, T. and Stephenson, G. J. Jr., (1984) Phys. Lett. **146 B**, 143;
—— Goldman, T. (1986) in *Quarks and Gluons in Particles and Nuclei*, edited by S. Brodsky and E. Moniz (World Scientific, Singapore), 363.

Govaerts, J., Mandula, J. E. and Weyers, J. (1984) Nucl. Phys. **B 237**, 59.

Griffin, J. J. and Wheeler, J. A. (1957) Phys. Rev. **108**, 311.

Gupta, S. R., Radford, S. F., and Repko, W. W. (1982) Phys. Rev. D **26**, 3305.

Haff, P. K. and Wilets, L. (1974) Phys. Rev. **C10**, 353.

Hansson, T. H., and Jaffe, R. L. (1983) Phys. Rev. D **28**, 882; Annals of Physics (N.Y.) **151**, 204.

Henyey, L. G., Wilets, L., Böhm, K. H., Le Levier, R. and Levee, R. D. (1959) Astrophys. J. **129**, 628.

Hill, D. L. and Wheeler, J. A. (1953) Phys. Rev. **89**, 1106.

Horn, R., Goldflam, R and Wilets, L. (1986) Comp. Phys. Com. **42**, 105.

Huang, K. and Stump, D. R. (1976) Phys. Rev. D **14**, 223.

Iwasaki, M. and Kondo, Y. (1987) Phys. Lett. **199 B**, 437.

Johnson, K. and Thorn, C. B. (1976) Phys. Rev **D13**, 1934.

Johnson, K. (1978) in *Current Trends in the Theory of Fields*, Proceedings, Edited by J. E. Lanutti and P. K. Williams (AIP, New York, 1978), 112.

Johnson, W. A., Howard, A. Q. and Dudley, D. G. (1979) Radio Science **14**, 961.

Kahana, S. and Ripka, G. (1984) Nucl. Phys. **A 429**, 462.

Kerman, A. K. and Dagdeviren, N. R. (1986) (private communication).

Köppel, Th. and Harvey, M. (1985) Phys. Rev. D **31**, 171.

Krajcik, R. A. and Foldy, L. L. (1974) Phys. Rev. **D 10**, 1777.

Krein, G., Tang, P., Wilets L. and Williams, A. G. (1988)
Phys. Lett. **B** (in press).

Krein, G., Tang, P., Wilets L. and Williams, A. G. (1989)
Private communication.

Krein, G., Tang, P. and Williams, A. G. (1988) Phys. Lett. **B** (in press).

Krein, G. and Williams, A. G. (1989), private communication.

Lee, T. D. and Wick, G. C. (1974) Phys. Rev. **D 9**, 2291.

Lee, T. D. (1979) Phys. Rev. **D 19**, 1802.

Lee, T. D. (1981) *Particle Physics and Introduction to Field Theory*
(Harwood Academic, New York).

Li, M. (1987) "Quantum Corrections to Solitons composed of
Interacting Fermions and Bosons," Doctoral Dissertation,
University of Washington.

Li, M., Birse, M. C. and Wilets, L. (1986) Journal of Physics **G 13**, 1.

Li, M., Perry, R. and Wilets, L. (1987) Phys. Rev. **D 36**, 596.

Li, M. and Perry, R. (1988) Phys. Rev. **D 37**, 1670.

Li, M., Wilets, L. and Perry, R. (1988) J. Comp. Phys. (to be published).

Lichtenberg, D. B. (1978) *Unitary Symmetry and Elementary Particles,* (2nd ed.)
(Academic Press, London).

Lübeck, E. G. (1986) "Momentum Projection and Revativistic Boost of
Solitons," Doctoral Dissertation, University of Washington;

—— Lübeck, E. G., Birse, M. C., Henley, E. M. and Wilets, L. (1986)
Phys. Rev **D 33**, 234.

Lübeck, E. G., Henley, E. M. and Wilets (1987) Phys. Rev. **D 35**, 2809.

Machleidt, R., Holinde, K. and Elster, Ch. (1987) Phys. Rep. **149**, 1.

Mandelstam, S. (1979) Phys. Rev. **D 19**, 2391.

McNeil, J. A., Shepard, R. J. and Wallace, S.J. (1983)
Phys. Rev. Lett., **50**, 1439.

Miller, G. A., Thomas, A. W. and Théberge, S. (1980) Phys. Lett. **91B**, 192;

—— Théberge, S., Thomas, A. W. and Miller, G. A. (1980)
Phys. Rev. **D 22**, 2838;

—— Thomas, A. W., Théberge, S. and Miller, G. A. (1981) Phys. Rev. **D 24**,
216.

Moxhay, P. and Rosner, J. R. (1983) Phys. Rev. **D 28**, 1132.

Nadkarni, S., Nielsen, H. B. and Zahed, I. (1985) Nucl. Phys. **B 253** 308.

—— Nadkarni, S. and Zahed, I. (1986) Nucl. Phys. **B 263** 23.

Nambu, Y. and Jona-Lasinio, G. (1961) Phys. Rev. **122**, 345; **124**, 246.

Nielsen, H. B. and Patkós, A. (1982) Nucl. Phys. **B 195**, 137.

Pagels, H. (1976) Phys. Rev. D **14**, 2747.

Peierls, R. E. and Yoccoz, J. (1957) Proc. Phys. Soc. London **A70**, 381.

Pirner, H. J., Wroldsen, J. and Ilgenfritz, M. (1987) Nucl. Phys. **B 294**, 905.

Pryce, M. H. L. (1948) Proc. R. Soc. London **A195**, 62.

Rafelski, J. (1977) Phys. Rev. **D 16**, 1890.

Reinhardt, H., Dang, B. V. and Schulz, H. (1985) Phys. Lett. **159 B**, 161.

Ring, P. and Schuck, P. (1980) *The Nuclear Many-body Problem* (Springer-Verlag, New York).

Roberts, C. D., Cahill, R. T. and Praschifka, J. (1988) Ann. Phys. (N. Y.) **187**.

Rosina, M., Schuh, A. and Pirner, H. J. (1986) Nucl. Phys. **A 448** , 557.

Russell, J. S. (1845) Pro. British Assoc. Advancement of Science (London), 311.

Saly, R. (1983) Comput. Phys. Commun. **30**, 411;

Saly, R. and Sundaresan, M. K. (1984) Phys. Rev. **D29**, 525.

Scadron, M. D. (1983) Ann. Phys. (N.Y.) **148**, 257.

Schuh, A. (1984) Private communication.

Schuh, A. (1985) 'Nukleon-Nukleon-Wechselwirkung im Soliton Bag Modell,' Doctoral Dissertation, University of Heidelberg;

— Schuh, A., Pirner, H. J. and Wilets, L. (1986) Physics Letters **B 174**, 10.

Schuh, A. and Pirner, H. J. (1986) Phys. Lett. **173 B**, 19. (1986) 19.

Serot, B. D. and Walecka, J. D. (1986) Adv. Nucl. Phys., **16**, ed. J.W. Negele and E. Vogt (Plenum, New York).

Shepard, R. J., McNeil, J. A. and Wallace, S. J. (1983) Phys. Rev. Lett., **50**, 1443.

Skyrme, T. H. R., (1961) Proc. Royal Soc. London **260A**, 127;

Stancu, F. and Wilets, L. (1987) Phys. Rev. C **36**, 726.

Stancu, F. and Wilets, L. (1988) Phys. Rev. C **38**, 1145.

— (1962) Nucl. Phys. **31**, 556.

Tang, P. and Wilets L. (1989) private communication.

ter Haar, B. and Malfliet, R. (1987) Phys. Rep. **149** 208.

Vasak, D. Wietschorke, K.-H., Müller, B. and Greiner, W. (1983) Zeit. Phys. **C 21**, 119.

Vento, V., Rho, M., Nyman, E. B., Jun, J. H., and Brown, G. E. (1980) Nucl. Phys. **A345**, 413.

Vepstas, L., Jackson, A. D. and Goldhaber, A. S. (1984) Phys. Lett. **140 B**, 280.

Vinciarelli, P. (1972) Nuovo Cimento Lett. **4**, 905.

Walecka, J. D. (1974) Ann. Phys. (N.Y.), **83**, 491.

Wampler, K. D. and Wilets, L. (1988) Computers in Physics **2**, 53.

Wichmann, E. H. and Kroll, N. M. (1956) Phys. Rev. **101**, 843.

Wilets, L. (1979) *Mesons in Nuclei*, Edited by M. Rho and D. Wilkinson (North Holland), 790.

Wilets, L. (1985) in *Hadrons and Heavy Ions*, Lecture Notes in Physics **231**, 317 (Springer, Berlin) .

Wilets, L. (1986) Foundations of Physics **16**, 171.

Wilets, L. and Berg, R, A. (1956) Phys. Rev. **101**, 201.

—- Wilets, L. (1958) Rev. Mod. Phys. **30**, 542.

Wilets, L., Birse, M. C., Lübeck, E. G., and Henley, E. M. (1984) Nucl. Phys. **A 434**, 129c;

—— Birse, M. C., Henley, E. M., Lübeck, G. and Wilets, L. (1984) in
   Solitons in Nuclear and Elementary Particle Physics,
   Proc. 1984 Lewes Workshop, eds. A. Chodos, E. Hadjimichael and H. C. Tze,
   (World Scientific, Singapore, 1984).
Williams, A. G., and Thomas, A. W. (1986) Phys. Rev. **C33**, 1070.
Witten, E. (1983) Nucl. Phys. **223B**, 422, 433;
—— Adkins, G. S., Nappi, C. R. and Witten, E. Nucl. Phys. **228B**, 552;
—— Adkins, G. S. and Nappi, C. R. (1984) Nucl. Phys. **233B**, 109.
Wosiek, J. and Haymaker, R. W. (1987) Phys. Rev. **D 36**, 3297.
Yang, C. N., and Mills, R. (1954) Phys. Rev **96**, 191.
Zabusky, N. J. and Kruskal, M. D. (1965) Phys. Rev. Letters **15**, 240.
Zhang, Q., Derreth, C., Schäfer A. and Greiner W. (1986)
   J. Phys. G **12**, L19.
Zweig, G. (1964) CERN Report No. 8182/Th 401; 8419/Th 412, unpublished.

INDEX